SpringerBriefs in Applied Sciences and Technology

T0214894

More information about this series at http://www.springer.com/series/8884

Axaykumar Mehta · Bijnan Bandyopadhyay

Frequency-Shaped and Observer-Based Discrete-time Sliding Mode Control

 Springer

Axaykumar Mehta
Institute of Infrastructure Technology
  Research and Management (IITRAM)
Ahmedabad, Gujarat
India

Bijnan Bandyopadhyay
IDP in Systems and Control Engineering
Indian Institute of Technology Bombay
Mumbai, Maharashtra
India

ISSN 2191-530X   ISSN 2191-5318   (electronic)
SpringerBriefs in Applied Sciences and Technology
ISBN 978-81-322-2237-8   ISBN 978-81-322-2238-5   (eBook)
DOI 10.1007/978-81-322-2238-5

Library of Congress Control Number: 2014957961

Springer New Delhi Heidelberg New York Dordrecht London

Printed on acid-free paper

Springer (India) Pvt. Ltd. is part of Springer Science+Business Media (www.springer.com)

*To our Parents, Teachers, Family and
Friends, who made us capable enough to
write this book*

Preface

It is well established that the sliding mode control strategy provides an effective and robust method of controlling the deterministic system due to its well-known invariance property to a class of bounded disturbance and parameter variations. Advances in microcomputer technologies have made digital control increasingly popular among researchers worldwide. This led to the study of discrete-time sliding mode control design and its implementation. However, most of the strategies proposed are based on the state feedback approach. But, system state variables are often not fully available for many practical systems. Also some state variables may be difficult/costly to measure and sometimes have no physical meaning and cannot be measured at all. Thus, one has to resort to output feedback method or observer-based design. Sliding mode control using static output feedback is not always possible. On the other hand observer-based design requires more hardware resources and also increases the dimension of the system. Recently, there have been efforts to design the multirate output feedback-based sliding mode control (MROFSMC) where the available output is measured at a faster rate than the input actuation rate and by means of that states are obtained implicitly. It has also been shown that sliding mode control using multirate output feedback is possible for all controllable/observable systems. In general, most of the methods available for sliding surface design are based on pole placement technique or Linear Quadratic (LQ) design. It does not take care of the certain frequency component excited/introduced during the sliding mode. So to suppress certain frequency dynamics during sliding mode the frequency shaped sliding mode may be used. The frequency shaped sliding mode control based on state feedback exists in the literature.

This monograph proposes a method for multirate frequency shaped sliding mode controller design based on switching and non-switching type of reaching law. In this approach, the frequency-dependent compensator dynamics are introduced through a frequency shaped sliding surface by assigning frequency-dependent weighing matrices in a linear quadratic regulator (LQR) design procedure. In this way the undesired high frequency dynamics or certain frequency disturbance can be eliminated. The states are implicitly obtained by measuring the output at a faster rate than the control input.

The vibration control of smart structure is a challenging problem as it has several vibratory modes. The frequency shaping approach may be used to suppress the frequency dynamics excited during sliding mode in smart structure. The frequency content of the optimal sliding mode is shaped by using a frequency-dependent compensator, such that a higher gain can be obtained at the resonance frequencies. The controllers based on the proposed method are designed and implemented for vibration suppression of the intelligent structure.

The monograph also presents the design of discrete-time reduced order observer using the duality to discrete-time sliding surface design. First, the duality between the coefficients of the discrete-time reduced order observer and the sliding surface design is established and then, the design method for the observer using Riccati equation is explained. Using the proposed method, the observer for the Power System Stabilizer (PSS) for Single Machine Infinite Bus (SMIB) system is designed and simulation is carried out using the observed states. The discrete-time sliding mode controller based on the proposed reduced order observer design method is also obtained for a laboratory experimental servo system and verified with the experimental results.

The monograph is organized in the following sequence.

In Chap. 1, the background of the sliding mode control strategy along with the research contributions of the monograph is discussed.

In Chap. 2, the preliminaries and literature survey of continuous-time sliding mode control and discrete-time sliding mode control along with the multirate output measurement technique for state estimation is presented.

In Chaps. 3 and 4, design method for multirate output feedback-based frequency shaped sliding mode control for vibration suppression of the intelligent structure with switching and non-switching type control law are proposed.

Chapter 5 presents the design of discrete-time reduced order observer using the duality to discrete-time sliding surface design along with the application for reduced order observer design for the Power System Stabilizer (PSS) of Single Machine Infinite Bus (SMIB) system and Industrial Servo System.

Keywords Variable structure control, Sliding mode control, Smart structure, Uncertain system, Duality, Power system stabilizer

Mumbai, India, October 2014   Axaykumar Mehta
                                                                        Bijnan Bandyopadhyay

Contents

Figures

Tables

Acronyms

Abbreviations

AVR   Automatic Voltage Regulator
DAC   Digital to Analog Convertor
DSMC   Discrete-Time Sliding Mode Control
FSSMC Frequency Shaped Sliding Mode Control
LTI   Linear Time Invariant
PSS   Power System Stabilizer
QSM   Quasi Sliding Mode
SMC   Sliding Mode Control
SMIB   Single Machine Infinite Bus
SR   Speed Reduction
VSC   Variable Structure Control

Symbols

A	State matrix in continuous-time model of LTI system
B	Input matrix in continuous-time model of LTI system
C	Output matrix
C_0	Measurement matrix of lifted system
D_0	Transmission matrix of lifted system
τ	Input sampling time
Δ	Output sampling time
N	Ratio of input and output sampling rates in a multirate system
ν	Observability index
Φ_τ	State (plant) matrix of the discrete system sampled at τ
Φ_Δ	State (plant) matrix of the discrete system sampled at Δ
Γ_τ	Control input matrix of the discrete system sampled at τ
Γ_Δ	Control input matrix of the discrete system sampled at Δ
L_y, L_u, L_d	Gains of multirate state estimation algorithm

$s(k)$	Sliding surface
C_s	Sliding gain
q, ρ	Tuning parameters of Gao's reaching law
d_k	Unknown matched disturbance
$s_d(k)$	Priori known function for Bartoszewicz's reaching law
J_{os}	Quadratic cost function
s_{os}	Optimal sliding function
Q	Positive definite matrix
J_{fs}	Frequency dependent weighing function
E_{fd}	Equivalent excitation voltage (field circuit voltage)
E'_d	Internal voltage behind transient reactance x'_q
E'_q	Internal voltage behind transient reactance x'_d
I_d, I_q	Direct and quadrature components of armature current
K_E, T_E	AVR gain and time constant, respectively
$K_1 - K_6$	Heffron-Philips constants
S_m	Slip
T'_{d0}	Open circuit transient time constant
T_m, T_e	Mechanical and electrical torques, respectively
V_d, V_q	Direct and quadrature components of terminal voltage
V_{ref}	Reference voltage
V_t	Generator terminal voltage
V_s	Correction voltage
x_d, x_q	Synchronous reactances in d and q axes, respectively
x'_d	Direct axis transient reactance
x'_q	Quadrature axis reactance
x_e	External reactance
X_s	Series reactance
δ	Rotor angle
ω	Angular speed

Authors' Biography

Axaykumar Mehta is born in Bharuch, Gujarat, India in 1975 and obtained his B.E. Electrical (1996), M.Tech (2002) and Ph.D. (2009) degrees from Gujarat University Ahmedabad, IIT Kharagpur and IIT Mumbai, respectively. He worked as Associate Faculty at Indian Institute Technology, Gandhinagar during 2010–2011. He also acted as Professor; Director at Gujarat Power Engineering and Research Institute, Mehsana, Gujarat, India during 2012–2014. Currently, he is Associate Professor at Institute of Infrastructure Technology Research and Management, Ahmedabad, Gujarat. His research interests include Nonlinear Sliding Mode Control and Observer, Sliding Mode Control Application in Electrical Engineering and Networked Control System. He has published 30 research papers in peer-reviewed international journals and conferences of repute. He is Senior Member IEEE, Life Member of Institution of Engineers (India), Life Member of Indian Society for Technical Education and Member of Systems Society of India. He was conferred the Best paper award by SSI and Pedagogical Innovation award 2014 by Gujarat Technological University.

Prof. Bijnan Bandyopadhyay received his B.E. degree in Electronics and Telecommunication Engineering from the University of Calcutta, Calcutta, India in 1978, and Ph.D. in Electrical Engineering from the Indian Institute of Technology, Delhi, India in 1986. In 1987, he joined the Interdisciplinary Programme in Systems and Control Engineering, Indian Institute of Technology Bombay, India, as a faculty member, where he is currently Professor. In 1996, he was with the

Lehrstuhl fur Elecktrische Steuerung und Regelung, Ruhr Universitat Bochum, Bochum, Germany, as an Alexander von Humboldt Fellow. He has been a visiting Professor at Okayama University, Japan, Korea Advance Institute Science and Technology (KAIST) South Korea and Chiba National University in 2007. He visited University of Western Australia, Australia as a Gledden Visiting Senior Fellow in 2007. Professor Bandyopadhyay is recipient of UKIERI (UK India Education and Research Initiative) Major Award in 2007, 'Distinguished Visiting Fellowship' award in 2009 and 2012 from "The Royal Academy of Engineering", London. Professor Bandyopadhyay is a Fellow of Indian National Academy of Engineering (INAE), Senior Member of IEEE and a Fellow of IETE (India). He has published 9 books and monographs, 6 book chapters and more than 300 journal articles and conference papers. He has guided 25 Ph.D. theses at IIT Bombay. His research interests include the areas of higher order sliding mode control, multirate output feedback control, discrete-time sliding mode control, large-scale systems, model order reduction, nuclear reactor control and smart structure control. Prof. Bandyopadhyay served as Co-chairman of the International Organization Committee and as Chairman of the Local Arrangements Committee for the IEEE International Conference in Industrial Technology, held in Goa, India, in Jan. 2000. He also served as one of the General Chairs of IEEE ICIT conference held in Mumbai, India in December 2006. Prof. Bandyopadhyay has served as General Chair for IEEE International Workshop on Variable Structure Systems held in Mumbai in January 2012.

Chapter 1
Introduction

Keywords Variable structure control · Discrete-time sliding mode control · Frequency shaping · Smart structure · Duality · Reduced order observer · Power system stabilizer · Industrial servo system

1.1 Background

The Sliding Mode Control (SMC) technique has basically evolved from the relay control theory. It is a well-known fact that the roots of the SMC technique are found in the then Soviet Union [10, 12] but it did not get published outside Soviet Union until a book by Itkis [21] and a paper by Utkin [38] were published. Since then significant work has been done on Variable Structure Control (VSC) and Sliding Mode Control, which may be refereed in the survey papers [1, 20, 46]. The most intriguing aspect of the sliding mode is the switching between two structures or feedback system such that a new motion called sliding motion occurs in a manifold. Thus the sliding motion arises in system due to discontinuous control and so the state trajectories are confined to the manifold in the state space. The design procedure consists of two stages: selection of the equation of the discontinuity surface such that the sliding mode in their intersection exhibits the desired dynamics, and then the selection of discontinuous control enforcing this motion in the designed manifold. This control strategy provides an effective and robust means of controlling the system due to its well-known invariance property to a class of bounded disturbance and parameter variations [9, 11, 39].

Traditionally, the term sliding mode is associated with discontinuities on the right-hand side of the equation of motion, since the phenomenon was revealed in continuous-time systems. But due to advances in microcomputer technologies, the digital control is increasingly popular among the researchers worldwide and this leads to study of Discrete-Time Sliding Mode Control (DSMC) design and its implementation. The concepts of discrete-time sliding mode and 'quasi sliding mode' were

© The Author(s) 2015
A. Mehta and B. Bandyopadhyay, *Frequency-Shaped and Observer-Based Discrete-time Sliding Mode Control*, SpringerBriefs in Applied Sciences and Technology, DOI 10.1007/978-81-322-2238-5_1

1

initially introduced by Milosavljevic [33] and the issues of stability of DSMC was discussed by Sarpturk et al. [36]. Since then many researchers have contributed to the DSMC theory [3, 13, 14, 25, 40].

The main difference between a Continuous-Time Sliding Mode Control (CSMC) and a Discrete-Time Sliding Mode Control (DSMC) is, apart from the model of the system under control, the implementation of the control law. A CSMC typically has a discontinuous control term which starts switching with infinite frequency once the system is driven to the sliding mode. While, in the case of DSMC, the measurement and control signal application are performed only at regular intervals of time and between these instants, the control signal is held constant so the switching frequency is finite. In DSMC, as the control input remains constant for the entire sampling period, the states can never be on the sliding surface but move in a zig-zag form called quasi-sliding mode motion [14]. There is also another school of thought for DSMC where the discontinuous control term may be eliminated from the control signal and so the control action becomes smooth and eliminates the chattering due to not having switching term [3].

It is also known that the design of the sliding surface plays a vital role in the design of sliding mode control and most of the methods available for sliding surface design are based on pole placement technique or Linear Quadratic (LQ) design. But the difficulty with these methods is that it does not take care of the certain frequency component excited/introduced during sliding mode. So, to suppress certain frequency dynamics during sliding mode the frequency shaped sliding mode may be used.

In this approach, the frequency-dependent compensator dynamics is introduced through a frequency shaped sliding surface by assigning frequency-dependent weighing matrices in a Linear Quadratic Regulator (LQR) design procedure so that the undesired high frequency dynamics or certain frequency disturbance can be eliminated. The frequency shaping approach to the LQ design was first proposed by Gupta [16] and the idea was further extended in papers [2, 4, 34, 45]. The idea is to transform the frequency domain performance to its equivalent time domain counterpart as proposed by Gupta [16], who considered the linear quadratic regulator problem using state feedback. The weighing matrices in the frequency shaped performance measure are expressed as a function of the frequency variable so that the system variables may be penalized over the specific band of frequencies. Anderson et al. [2] investigated the problem of selection of frequency-dependent weighing factor and its effect on robustness of the closed loop system. Moore and Minigori [34] discussed frequency shaped LQ design and spectral factorization. Young and Ozguner [45] considered only the frequency-dependent control weighing matrix. While in [26], all the possible cases for which the weighing function may be frequency dependent are studied. Xu and Cao [44] proposed a synthesis method of sliding mode controller using frequency shaping and terminal attractor methods for single link r-mode robotic manipulator and also proved the asymptotical stability of the synthesized sliding mode by Popov's criterion. The discrete-time approach to the frequency shaping LQ control is proposed in [5] where the discrete-time LQ optimal control for multi-variable system is developed using the discrete-time Parseval's Theorem.

Recently, Wu and Liu [43] proposed discrete-time frequency shaped sliding mode control law for flying height control of near-field optical disk drives.

In all the above work, it is mandatory to have the state information for the robust control. But in many situations, the states are seldom available for measurement and so the observer must be designed to get the state information. The Luenberger observer may be used for the state estimation but it introduces additional dynamics in the closed loop. It also increases the overall system dimension and complexity of the system. Instead, we may use the multirate output measurement technique [22, 41] for estimation of the states where the error in the states go to zero in one sample time. Moreover, the method does not increase the overall dimension of the system.

Multirate sampled-data system is a digital control system where each input and output is updated at a different sampling rate. Research in multirate sampled-data system has evolved since the paper published by Krank [24, 27] but received considerable attention due to papers by various authors [17, 41, 42]. The multirate sampling is often required and also beneficial due to the technological constraints and economic benefits. Furthermore, properly designed control systems using multirate sampled-data mechanism can significantly improve the performance of the controlled system [6]. Recently, there have been efforts to design the Multirate Output Feedback based Sliding Mode Control (MROFSMC) where the available output is measured at a faster rate than the input actuation rate and by means of that the states are obtained implicitly [22]. It has also been shown that sliding mode control using multirate output feedback is possible for all controllable/observable systems [23].

In this monograph, the design method of multirate output feedback-based frequency shaped sliding mode control with switching and nonswitching type reaching laws are proposed [28, 31] and used for vibration control of smart structure. The frequency shaping approach may be used to suppress certain frequency dynamics during sliding mode in smart structure.

The vibration control of smart structure is a challenging problem as it has several vibratory modes. The concept of the smart or intelligent structure is given in [35]. A smart structure typically consists of a host structure with sensors and actuators coordinated by a controller. The integrated system is called a smart structure because it has the ability to perform self-diagnosis and adapt to the change in environment. The technology of smart materials and structures especially piezoelectric smart structures has become mature over the last decade. One promising application of piezoelectric smart structures is control and suppression of unwanted structural vibrations [37]. Fully active actuators like electromagnetic shakers, piezoelectric ceramic and film, magnetostrictive, and electro-hydraulic devices can be used to generate a secondary vibrational response in a linear mechanical system which could reduce the overall response by destructive interference with the original response of the system caused by the primary source of vibration [7, 19]. In [8] the concept of smart structure and its applications is discussed at length. Hubbard and Baily [19] studied the application of piezoelectric material as sensor/actuator. Hanagud et al. [18] developed Finite Element Method (FEM) model for an active beam with many distributed piezoceramic sensor/actuator. Feedback control of vibration in mechanical flexible system has application to robotics, aircraft, space structure, and in many industrial problems. The controllers based on proposed methods are designed and implemented for vibration suppression of the intelligent structure [30].

In some situations, the observer-based method may be preferred over multirate output feedback due to presence of noise or sampling limitation. Hence, a considerable amount of research has been carried out on observer and several authors have proposed several design method for the observers. Among all the methods, the pole placement technique and the optimal observer design [11] are widely used methods. Both the methods require the observer system to be transformed into Gopinath's form [15]. But when the observer system is transformed to Gopinath's form, the original controllable canonical form is lost. The duality between discrete-time sliding surface design and discrete-time reduced order observer (Reduced Order Luenberger's Observer) may be used to develop the observer-based sliding mode control strategy.

In this monograph, a method for designing a reduced order observer based on duality principle is discussed and the same is used for Power System Stabilizer and Industrial Emulator Servo System [29, 32].

1.2 Contributions of the Monograph

The main contributions discussed in this monograph are summarized as follows.

1.2.1 Multirate Output Feedback-Based Frequency Shaped Sliding Mode Control: A Switching Type Control Law

The third chapter of this monograph discusses a design method for multirate output feedback-based frequency shaped sliding mode control [28]. A compensator is introduced in the sliding surface to attenuate the certain frequency component of the sliding mode dynamics. The switching type reaching law is used for designing the control law and the states are estimated from the multirate output measurements. The algorithm is applied for vibration control of the smart structure when the resonance modes are excited by external disturbance.

1.2.2 Multirate Output Feedback Based Frequency Shaped Sliding Mode Control: A Nonswitching Type Control Law

A design method for multirate output feedback-based frequency shaped sliding mode control with nonswitching type control law is presented in Chap. 4 [30, 31]. The method improves the performance of the control algorithm significantly while applied to the vibration control of the smart structure excited at resonance frequency. The algorithm is more practical for implementation.

1.2.3 Reduced Order Observer Design Using Duality to Sliding Surface Design for Sliding Mode Control

In Chap. 5, the design method of discrete-time reduced order observer using duality to discrete-time sliding surface design is presented [32]. First, the duality between the coefficients of the discrete-time reduced order observer and the sliding surface design is established and then, the design method for the observer using the Riccati equation is explained. Using the duality between the coefficients of the reduced order observer and the sliding surface, a reduced order observer is designed for the Power System Stabilizer (PSS) of Single Machine Infinite Bus (SMIB) system [29]. The simulation results are presented to show the efficacy of the PSS with the reduced order observer designed by the proposed duality. It enhances the stability and also improves the dynamic response of the (SMIB) operating in different conditions. Also a discrete-time sliding mode controller based on the proposed observer method is designed and tested on a laboratory experimental servo system [32].

References

1. Adamy J, Flemming A (2004) Soft variable-structure controls: a survey. Automatica 40:1821–1844
2. Anderson BDO, Moore JB, Mongori DL (1987) Relations between frequency dependent control and state weighing in LQG problems. Optim Control Appl Methods 08:109–127
3. Bartoszewicz A (1998) Discrete-time quasi sliding mode control strategies. IEEE Trans Ind Electron 45(04):633–637
4. Cheok KC, Hu HX, Loh NK (1988) Optimal output feedback regulation with frequency shaped cost functionals. J Guid Control 47(06):1665–1681
5. Cheok KC, Hu HX, Loh NK (1989) Discrete time frequency shaping parametric LQ control with application to active seat suspension control. IEEE Trans Ind Electron 36(03):383–390
6. Colaneri P, Scattolini R, Schiavoni N (1990) Stabilization of multi-rate sampled-data linear systems. Automatica 33(26):377–380
7. Crawley EF, Luis JD (1987) Use of piezoelectric actuators as elements of intelligent structures. AIAA J 25:1373–1385
8. Culshaw B (1992) Smart structure a concept or reality. J Syst Control Eng 26(206):01–08
9. Drakunov SV, Utkin VI (1990) Sliding mode in dynamic systems. Int J Control 55:1029–1037
10. Drazenovic B (1969) The invariance conditions in variable structure systems. Automatica 05:287–295
11. Edwards C, Spurgeon SK (1998) Sliding mode control: theory and applications. Taylor and Francis, London
12. Emelyanov SV (1967) Variable structure control system (In Russian). Nauka, Moscow
13. Furuta K (1990) Sliding mode control of a discrete system. IEEE Syst Control Lett 14:144–152
14. Gao W, Wang Y, Homaifa A (1995) Discrete-time variable structure control system. IEEE Trans Ind Electron 42(02):117–122
15. Gopinath G (1999) On the control of linear multiple input-output system. Bell Syst Technol J 50(01):50–55
16. Gupta NK (1985) Frequency shaped cost functionals: extension of linear quadratic-Gaussian design methods. J Guid Control 03(06):1665–1681
17. Hagiwara T, Araki M (1988) Design of a stable state feedback controller based on the multi-rate sampling of plant output. IEEE Trans Autom Control 33(09):812–819

18. Hanagud S et al (1992) Optimal vibration control by the use of piezoelectric sensors and actuators. J Guid, Control Dyn 15(05):1199–1206
19. Hubbard JEJ, Baily T (1985) Distributed Piezoelectric polymer active vibration control of a cantilever beam. J Guid, Dyn Control 08(05):605–611
20. Hung JY, Gao W, Hung JC (1993) Variable structure control: a survey. IEEE Trans Autom Control 40(01):2–22
21. Itkis U (1976) Control systems of variable structure. Wiley, New York
22. Janardhanan S, Bandyopadhyay B (2006) Output feedback sliding mode control for uncertain systems using fast output sampling technique. IEEE Trans Ind Electron 53(05):1677–1682
23. Janardhanan S, Bandyopadhyay B (2007) Multirate output feedback based robust quasi-sliding mode control of discrete-time systems. IEEE Trans Autom Control 52(03):499–503
24. Jury E (1967) A note on multirate sampled-data systems. IRE Trans Autom Control 12(03):319–320
25. Kaynak O, Denker A (1993) Discrete-time sliding mode control in the presence of system uncertainty. Int J Control 57(05):1177–1189
26. Koshkouei AJ, Zinober ASI (2000) Robust frequency shaping sliding mode control. IEE Proc Control Theory Appl 147:312–320
27. Kranc GM (1957) InputOutput analysis of multi-rate feedback system. IRE Trans Autom Control 03:21–28
28. Mehta A, Bandyopadhyay B (2006) Multirate output feedback based frequency shaped sliding mode control. In: Proceedings of IEEE international conference on industrial technology (ICIT2006), Mumbai, India, pp 2658–2662
29. Mehta A, Bandyopadhyay B (2007) Reduced order observer design for power system stabilizer using the duality to discrete time sliding surface design. In: Proceedings of 33rd annual conference of the IEEE industrial electronics society (IECON2007), Taipei, Taiwan, pp 908–914
30. Mehta A, Bandyopadhyay B (2009) Frequency-shaped sliding mode control using output sampled measurements. IEEE Trans Ind Electron 56(01):28–35
31. Mehta A, Bandyopadhyay B (2010) The design and implementation of output feedback based frequency shaped sliding mode controller for the smart structure. In: Proceedings of 2010 IEEE symposium on industrial electronics (ISIE2010), Bari, Italy, pp 353–358
32. Mehta A, Bandyopadhyay B, Inoue A (2010) Reduced-order observer design for servo system using duality to discrete-time sliding-surface design. IEEE Trans Ind Electron 57(11):3793–3800
33. Milosavljevic C (1985) General conditions for the existence of quasi sliding mode on the switching hyperplane in discrete variable structure systems. Autom Remote Control 46:307–314
34. Moore JB, Minigori DL (1988) Robust frequency shaped LQ control. Automatica 23:641–646
35. Rogers CA (1992) Intelligent material systems—the dawn of a new materials age. Sci Mach J 46(09):977–983
36. Sarpturk SZ, Istefanopulos Y, Kaynak O (1987) On the stability of discrete-time sliding mode control systems. IEEE Trans Control Syst 32(10):930–932
37. Sunar M, Rao SS (1999) Recent advances in sensing and control of flexible structure via piezoelectric materials technology. ASME Appl Mech Rev 52:01–16
38. Utkin VI (1977) Variable structure systems with sliding mode. IEEE Trans Autom Control 22(02):212–222
39. Utkin VI (1992) Sliding modes in control optimization. Springer, New York
40. Utkin VI, Drakunov SV (1989) On Discrete-time sliding modes control. In: Proceedings of IFAC conference on nonlinear control, Capri, Italy, pp 484–489
41. Werner H (1998) Multimodel robust control by fast output sampling—an LMI approach. Automatica 34(12):1625–1630
42. Werner H, Furuta K (1995) Simultaneous stabilization by piecewise constant periodic output feedback. Control Theory Adv Technol 10(04):1763–1775
43. Wu WC, Liu TS (2005) Frequency shaped sliding mode control for flying height of pickup head in near field optical disc drives. IEEE Trans Magn 41(02):1061–1063

44. Xu JX, Cao WJ (2000) Synthesized sliding mode control of a single-link flexible robot. Int J Control 73(03):197–209
45. Young KD, Ozguner U (1993) Frequency shaping compensator design for sliding mode. Int J Control 57(05):1005–1019
46. Young KD, Utkin VI, Ozguner U (1999) A control engineers guide to sliding mode control. IEEE Trans Autom Control 07(03):328–342

Chapter 2
Preliminaries of Sliding Mode Control

Abstract In this chapter, a brief introduction of the concept of sliding mode control and the evolution of discrete-time sliding mode from the continuous-time sliding mode is discussed. The concept of the multirate output feedback-based sliding mode control technique is also discussed.

Keywords Discrete-time sliding mode control · Output feedback · Reaching law · Multirate output feedback

2.1 Variable Structure Control

Variable Structure Control (VSC) with Sliding Mode Control (SMC) was first presented and elaborated in the 1960s in the then Soviet Union by Emelyanov [9, 13] and other researchers [20, 34]. Since then, VSC has been developed into a general design method being examined for a wide spectrum of system types including nonlinear systems, multi-input–multi-output systems, large scale systems, infinite dimensional systems, and stochastic systems. Also, the objectives of VSC have been extended from stabilization to other control functions. The main feature of the VSC is the invariance to a class of bounded disturbance and parameter variations [10, 34]. In Variable Structure Systems, the system is assumed to consist of continuous subsystems known as structures. These structures are changed or switched depending on the state of the system. The gain of a system may be changed or the transfer function of the system may be completely changed in these types of systems. The times (states) at which the structures change contribute to discontinuity surfaces in the phase planes. These surfaces are called as switching surfaces. If the switching surface satisfies the condition of having positive attraction, then such a surface would become a sliding surface [35].

A simple example of such a variable structure system would be a second-order system having system equations

$$\dot{x}_1 = x_2$$
$$\dot{x}_2 = ax_1 + bx_2 + u \qquad (2.1)$$

A. Mehta and B. Bandyopadhyay, *Frequency-Shaped and Observer-Based Discrete-time Sliding Mode Control*, SpringerBriefs in Applied Sciences and Technology, DOI 10.1007/978-81-322-2238-5_2

where x_1, x_2 are system states and a, b are system parameters. The system has feedback input given by

$$u = -\Psi x_1$$

The parameter Ψ is a variable parameter that takes values α and β as the structure changes. Suppose the system with input as α has complex eigenvalues with positive real part and the system with input as β has eigenvalues real but one positive and one negative, then the system trajectories in the two structures are both unstable as shown in Fig. 2.1. The complex eigenvalues give an unstable focus, whereas the one positive and one negative real eigenvalue gives a saddle point.

If we observe the phase portrait carefully, we can notice that the two unstable structures have certain regions of stability, like the describing point moves toward the saddle point along the eigenvector corresponding to the negative eigenvalue. To have the desired regions of the two structures in the resultant system, two switching surfaces are selected.

$$x_1 = 0 \tag{2.2}$$

$$s = cx_1 + x_2 = 0 \tag{2.3}$$

Selecting the switching law from these two equations, we get

$$\Psi = \begin{cases} \alpha, & \text{when } x_1 s > 0 \\ \beta, & \text{when } x_1 s < 0 \end{cases}$$

The phase portrait of the resultant system is as shown in Fig. 2.2. As we can see the switching surface $x_1 = 0$ has attraction properties only on one side of the surface, so no sliding occurs. But the switching surface s has attraction property on both sides of the surface, as a result this surface becomes the sliding surface of the system. If we look at the resultant motion on the sliding surface, the describing point slides toward the equilibrium point and hence the closed-loop system is stable. A variable

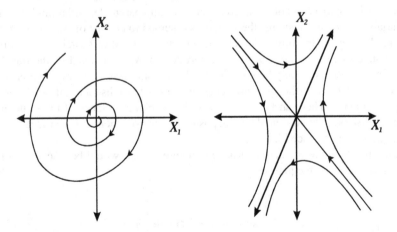

Fig. 2.1 Phase portrait of two unstable structure

Fig. 2.2 Phase portrait of
switched structure

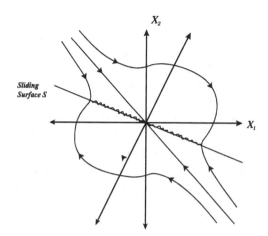

structure system consists of a set of continuous subsystems with a proper switching
logic and, as a result, control actions are discontinuous function of system state,
disturbance, and reference input. The sliding mode is the principal mode in variable
structure systems.

Definition 2.1 Sliding Mode: It is the motion of the system trajectory along a chosen
line/plane/surface of the state space.

2.2 Continuous-Time Sliding Mode Control

The sliding mode control can be viewed as a control process consisting of two
important phases:

- The reaching phase: The reaching phase is the part where the describing point starts
  from its initial condition and moves toward the sliding surface. During this period,
  however, the tracking error cannot be controlled directly and the system response is
  sensitive to parameter variations and noise. Thus, one would ideally like to shorten
  the duration or even eliminate the reaching phase. One easy way to minimize the
  reaching phase and hence the reaching time is to employ a larger control input.
  This, however, may cause extreme system sensitivity to unmodelled dynamics,
  actuator saturation, and undesirably higher chattering as well. The robustness of
  the VSC can be improved by shortening the reaching phase or may be guaranteed
  during the whole intervals of control action by eliminating the reaching phase.
  Several methods have been reported in the literature to eliminate the reaching
  phase completely [7, 8, 24, 30, 38]. The algorithms proposed in these papers
  employed piecewise constant sliding lines, i.e., the lines which move step by
  step and not continuously. This causes existence of many short reaching phases
  after each instant when the line is moved and does not allow to ensure system

insensitivity. This issue was resolved by Bartoszevicz [1] where the piecewise constant lines were replaced with continuously time-varying lines which indeed eliminate the reaching phase. The concept is applied further to several systems [3].

- The sliding phase: This is the phase in which the describing point moves only on the desired sliding surface. In this phase, the describing point does not necessarily follow any system trajectory that was present in the original fixed input system. This is because at the sliding surface the input continuously switches, and the system description is essentially discontinuous.

To find the equation of the system along the sliding surface many methods have been proposed. This is due to the fact that the differential equation has a nonanalytic right-hand side, which is the relay-type discontinuity. Consider a nth order system represented in the phase variable form

$$\dot{x}_i = x_{i+1}, \quad i = 1, 2, \ldots, n-1 \tag{2.4}$$

$$\dot{x}_n = -a_n x_n + \cdots + a_1 x_1 + Bu \tag{2.5}$$

The sliding surface is defined as

$$s(t) = C_s x(t). \tag{2.6}$$

The vector C_s consist of coefficients that describe the sliding surface in terms of the state vector $x(t)$. The sliding surface defined in such a way is called a hyperplane. The surface need not be a plane (or line in case of second-order system) always, the surface can be of any shape. In that case, the vector C_s is the gradient of the sliding surface, let us say G. If the sliding surface is a plane, then the gradient of the matrix is the matrix itself. The value of s specifies the distance of the point from the sliding surface, hence $s = 0$ implies the point that is on the sliding surface.

Defining the sliding surface as

$$s = c_{s1} x_1 + c_{s2} x_2 + \cdots + c_{sn-1} x_{n-1} + x_n = 0 \tag{2.7}$$

$$x_n = -c_{s1} x_{s1} - c_{s2} x_2 - \cdots - c_{sn-1} x_{n-1} \tag{2.8}$$

$$\dot{x}_n = -c_{s1} x_2 - c_{s2} x_3 - \cdots - c_{sn-2} x_{n-1} + \sum_{i=1}^{n-1} c_{sn-1} c_{si} x_i. \tag{2.9}$$

Thus, the entire dynamics of the system is governed by the sliding line/surface parameters only.

$$\dot{x}_i = x_{i+1}, \quad i = 1, 2, \ldots, n-1 \tag{2.10}$$

$$\dot{x}_n = -c_{s1} x_2 - c_{s2} x_3 - \cdots - c_{sn-2} x_{n-1} + \sum_{i=1}^{n-1} c_{sn-1} c_{si} x_i. \tag{2.11}$$

The system dynamics are independent of system parameters and determined by the surface parameters C_s only.

At the outset, two important properties are achieved during the sliding motion, that is, robustness and order reduction. However, to induce the sliding mode the following properties should exist:

- The system stability confined to sliding surface;
- Sliding mode should start in finite time.

The condition for first requirement is obtained as given below. Consider the system in regular form as

$$\dot{x}_1 = A_{11}x_1 + A_{12}x_2 \tag{2.12}$$

$$\dot{x}_2 = A_{21}x_1 + A_{22}x_2 + Bu \tag{2.13}$$

If the sliding surface is designed as

$$s = \begin{bmatrix} k & 1 \end{bmatrix} \begin{bmatrix} x_1 \\ x_2 \end{bmatrix}, \tag{2.14}$$

then the system dynamics confined to the sliding surface $s = kx_1 + x_2 = 0$ is given by

$$\dot{x}_1 = A_{11}x_1 + A_{12}x_2 = (A_{11} - A_{12}k)x_1. \tag{2.15}$$

If k is so designed that $A_{11} - A_{12}k$ has eigenvalues on LHP only, then the dynamics of x_1 is stable. Since $kx_1 + x_2 = 0$, the dynamics of x_2 is also stable. Hence, if the sliding surface is designed as $s = C_s x = kx_1 + x_2$, then the system dynamics confined to $s = 0$ is stable.

The second requirement is that the sliding mode should start in finite time. In the sliding phase, the describing point is supposed to move along the chosen surface. This in turn dictates that the sliding surface should be such that it has on both sides state trajectories corresponding to the two structures coming into it. If s is the distance of the describing point from the surface, then positive value of s implies that the point is above the sliding surface, whereas a negative value of s implies the point is below the sliding surface. $\dot{s}$ is the rate of change of distance from the sliding surface. Hence for the sliding motion to exist on the surface, the condition that needs to be satisfied is

$$s\dot{s} < 0. \tag{2.16}$$

This is called the 'reachability condition'. Next, it is shown here that the reachability condition is also not sufficient for the sliding. To show that, consider the example

$$\dot{s} = -s, \tag{2.17}$$

$$s\dot{s} = -s^2, \forall s \neq 0. \tag{2.18}$$

For which the solution for $s(t)$ is given by

$$s(t) = e^{-t}s(0), \tag{2.19}$$

that gives $s(t) = 0$ as $t \to \infty$. So, it takes infinite time to reach on the surface as it approaches the surface. To overcome the situation another condition is defined as

$$s\dot{s} < -\eta|s|, \quad \eta > 0 \tag{2.20}$$

This condition is known as 'η-*reachability condition*' that defines the minimum rate of convergence.

2.2.1 Reaching Law Approach

In the reaching law approach, the dynamics of the sliding function is directly expressed. Let the dynamics of the switching function be specified by the differential equation

$$\dot{s} = -qf(s) - k\text{sgn}(s) \tag{2.21}$$

$$q, k > 0 \tag{2.22}$$

$$sf(s) > 0, \quad \forall s \neq 0. \tag{2.23}$$

The control law may be obtained directly by the condition $\dot{s} = 0$ for the system

$$\dot{x}(t) = Ax(t) + Bu(t), \tag{2.24}$$

$$y(t) = Cx(t), \tag{2.25}$$

as

$$u(t) = -(CB)^{-1}(CAx(t) + qf(s) + k\text{sgn}(Cx(t))) \tag{2.26}$$

Similarly, the other reaching laws proposed in the literature are

- Constant rate reaching law
$$\dot{s} = -k\text{sgn}(s)$$

- Constant—proportional rate

$$\dot{s} = -qs - k\text{sgn}(s)$$

- Power-rate reaching law

$$\dot{s} = -k|s|^\alpha \text{sgn}(s), \quad 0 < \alpha < 1$$

2.2.2 Fillipov's Condition

The VSS dynamics is characterized by differential equation with discontinuous right-hand side. Filippov [15] first gave the solution for this type of system. The resultant

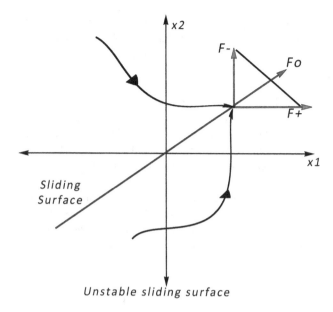

Fig. 2.3 Fillipov condition

direction of motion of the describing point along the sliding surface is specified by
the Fillipov's Vector (F_0). According to Fillipov if the state motion vector on one
side of the sliding surface is F^+ and on the other side of the surface is F^-, then the
resultant vector is given by the convex sum of the two vectors

$$F_0 = \mu F^+ + (1 - \mu)F^-,$$

where $0 < \mu < 1$.

The parameter μ depends on the magnitudes and directions of the vectors F^+, F^-
and the gradient of the sliding surface s. In the case of n vectors, this condition is
generalized to linear combination of all the vectors where the sum of coefficients of
the combination is unity (Fig. 2.3).

2.2.3 Limitations of Continuous-Time Sliding Mode Control

Variable structure control systems are high-speed switching feedback control systems,
which are known to be insensitive to matched uncertainties [9]. However, unmatched
uncertainties in physical systems may be present and may destroy the stability of the
sliding mode. In other words, if the invariance condition (matching condition) is
not satisfied, unmatched uncertainties will enter into the dynamics of the system in
the sliding mode. Thus, the system behavior in the sliding mode is *not invariant
to unmatched uncertainties*. Another obstacle for sliding mode to become useful

in practical systems is the high-frequency switching which results in '*chattering phenomenon*'. One of the causes for the chattering phenomenon is the presence of finite time delays for control computation and finite delay in switching. In the absence of switching delays, the switching device switches ideally at an infinite frequency. The second cause is the limitations of physical actuators and sensors, whose dynamics are often neglected. These parasitic dynamics in series with the plant cause small amplitude high-frequency nondecaying oscillations to appear in the neighborhood of the sliding manifold [19]. These oscillations are also referred as chattering. They excite the unmodeled high-frequency dynamics of the system. The chattering is not preferable from a practical point of view because it results in low control accuracy, high heat losses in electrical power circuits, and high wear of moving mechanical parts. Thus, the controller with a high switching frequency will cause fatigue of plant and reduce the service life of a machine. Several solutions are proposed for the reduction of chattering. One of the approaches to reduce chattering is to replace the relay control by saturating continuous approximation [19, 39].

2.3 Discrete-Time Sliding Mode Control

In case of continuous-time sliding mode, once the closed-loop system is driven into the sliding mode, a discontinuous control term switches with infinite frequency and that makes the main difference between a CSMC and a DSMC. The DSMC is automatically constrained to the sampling frequency due to limited sampling frequency. It means the control signal inevitably changes at the sample instances only. Moreover, in DSMC the control input remains constant for the entire sampling period. So, the states can never be on the sliding surface and move in zigzag form called quasi-sliding mode motion [16]. Thus, DSMC does not possess the invariance property which is found in CSMC as the invariance property is achieved only when the system states are exactly on the sliding surface. The robustness issues in DSMC are also still under investigation.

The concept of discrete-time sliding mode was first introduced by Milosavljevic [28] and further extended by Utkin and Drakunow [36]. Since then much work has been done in the field and many new algorithms are proposed.

Similar to CSMC, the design procedure for DSMC includes two steps:

- Computation of sliding surface

$$s(k) = C_s x(k) \tag{2.27}$$

  which has stable internal dynamics and
- Establishing a control law which steers the closed-loop system toward the sliding surface and ensures the system trajectories to stay as close as possible to the surface.

The first step of the design procedure is exactly the same as the design procedure presented for the CSMC. It is assumed that the closed-loop system is kept close enough to the sliding surface to approximate the switching function $s(k)$ by zero. The second step of the design procedure is different for DSMC as compared to CSMC in case of reaching law approach and is presented in the following section. The main difference is in the definition of a reaching law that is not as straightforward as for the continuous-time case.

2.3.1 State-Based Discrete-Time Sliding Mode Control

Consider the continuous-time linear time-invariant system

$$\dot{x}(t) = Ax(t) + Bu(t) + Ef(t), \tag{2.28}$$
$$y(t) = Cx(t), \tag{2.29}$$

where $x \in \Re^n$ is the state variable, $A \in \Re^{n \times n}$, $B \in \Re^{n \times m}$ is full rank, $u \in \Re^m$ is the control input, $C \in \Re^{p \times n}$ such that CB is nonsingular and $y \in \Re^p$ is the output. We assume that (A, B) is completely controllable and $m < n$. Let, the system in Eqs. (2.28) and (2.29) be discretized at τ sampling instant, then the discrete-time system is given by

$$x(k+1) = \Phi_\tau x(k) + \Gamma_\tau u(k) + E_\tau f(k), \tag{2.30}$$
$$y(k) = C_\tau x(k). \tag{2.31}$$

Representing the system in regular form as

$$\begin{bmatrix} x_1(k+1) \\ x_2(k+1) \end{bmatrix} = \begin{bmatrix} \Phi_{11} & \Phi_{12} \\ \Phi_{21} & \Phi_{22} \end{bmatrix} \begin{bmatrix} x_1(k) \\ x_2(k) \end{bmatrix} + \begin{bmatrix} 0 \\ \Gamma_2 \end{bmatrix} u(k) + \begin{bmatrix} 0 \\ E_{\tau 2} \end{bmatrix} f(k) \tag{2.32}$$

with the assumption that the uncertainty changes at the sampling instant only. Various sliding mode control laws have been proposed for the discrete-time system using different reaching laws as given below.

- Sarpturk's Reaching Law: Sarpturk [31] presented a reaching law which is direct discretized version of continuous-time sliding mode is given by

$$|s(k+1)| < |s(k)|.$$

Here, the sliding function is always directed toward the surface and also the norm of $s(k)$ monotonically decreases. The reaching law may be written in another way as

$$(s(k+1) - s(k))\mathrm{sgn}(s(k)) < 0 \tag{2.33}$$
$$(s(k+1) + s(k))\mathrm{sgn}(s(k)) > 0 \tag{2.34}$$

The first condition implies that the closed-loop system should be moving in the direction of the sliding surface, whereas the second condition implies that the closed-loop system is not allowed to go too far in that direction. In other words, the condition in Eq. (2.33) results in a lower bound for the control action and the same in Eq. (2.34) results in an upper bound. In [31], following control law is proposed

$$u(k) = -k(x, s)x(k),$$ (2.35)

where the gain k is given by

$$k(x, s) = \begin{cases} k^+, & \text{when } x(k)s(k) > 0 \\ k^-, & \text{when } x(k)s(k) < 0 \end{cases}$$

The computation of the coefficients k^+ and k^- is not an easy task. They can be determined by evaluating the conditions (2.33) and (2.34) resulting in an upper and a lower bound for each k^+ and k^-. Indeed, there are circumstances where they do not exist at all.

- Gao's Reaching Law: In order to design a DSMC, the Gao's reaching law [16] is adopted as

$$s(k + 1) = (1 - q\tau)s(k) - \rho\tau\text{sgn}(s(k))$$ (2.36)

where, $\tau > 0$ is the sampling time, $q > 0$, $\rho > 0$, and $1 - q\tau > 0$.
The DSMC is required to achieve the following performances [16].

1. Starting from any initial state, the trajectory will move monotonically toward the switching plane and cross it in finite time.
2. Once the trajectory has crossed the switching plane, it will cross the plane again in every successive sampling period, resulting in a zigzag motion about the switching plane.
3. The size of each successive zigzag step is nonincreasing and the trajectory stays within a specified band.

The control law with above reaching law (2.36) is derived for the system (2.30) as

$$u(k) = -(C_s \Gamma_\tau)^{-1}[C_s \Phi_\tau x(k) - (1 - q\tau)s(k) + \rho\tau\text{sgn}(s(k))]$$ (2.37)

The magnitude δ_s of quasi-sliding mode band (QSMB) for $s(k)$ that achieves the DSMC performance (2) can be computed by solving Eq. (2.36) for $s(k) = \delta_s$ and $s(k + 1) = -s(k)$. So

$$-2\delta_s = -q\tau\delta - \rho\tau$$

That gives,

$$\delta_s = \frac{\rho\tau}{2 - q\tau}$$ (2.38)

The control law (2.37) has two parameters ρ and q for tuning the response. From (2.38), ρ is directly proportional to the QSMB, and the system will overshoot when ρ is too large. On the other hand, large ρ could speed up transient response. From

(2.37), $q\tau$ is required to be smaller than one, so q has to be smaller than $1/\tau$, but large q could speed up transient response.

- Bartoszewicz's Reaching Law: Bartoszewicz [2] proposed a reaching law as

$$s(k+1) = d(k) - d_0 + s_d(k+1), \tag{2.39}$$

where the unknown $d(k)$ is defined as $d_l \leq d(k) = C_s^T \Delta\Phi_\tau x(k) + C_s^T E_\tau f(k) \leq d_u$ with d_l as lower bound and d_u as upper bound. Also d_0 and δ_d is given by

$$d_0 = \frac{d_l + d_u}{2} \text{ and } \delta_d = \frac{d_u - d_l}{2}.$$

$s_d(k)$ is an a priori known function such that the following applies:

- If $s(0) > 2\delta_d$ then

$$s_d(0) = s(0) \tag{2.40}$$
$$s_d(k) \cdot s_d(0) \geq 0 \quad \text{for any } k \geq 0 \tag{2.41}$$
$$s_d(k) \geq 0 \quad \text{for any } k \geq k^* \tag{2.42}$$
$$|s_d(k+1)| < |s_d(k)| - 2\delta_d \quad \text{for any } k \leq k^* \tag{2.43}$$

The above relations state that the time-dependent hyperplane monotonically, and in a finite time, converges from its initial position to the origin of the state space. Furthermore, in each control step, the hyperplane moves by the distance greater than $2\delta_d$. This, together with (2.39), implies that the reaching condition is satisfied, even in the case of the worst combination of disturbance in any two consecutive time steps.

- Otherwise, i.e., if $s(0) < 2\delta_d$ then $s_d(k) = 0$ for any $k \geq 0$.

The constant k^* in the above relations, is a positive integer chosen by the designer in order to achieve good trade off between the fast convergence rate of the system and the magnitude of the control required to achieve this convergence rate. The control law that satisfies the reaching law in Eq. (2.39) can be computed for system in Eq. (2.30) as

$$u(k) = -(C_s^T \Gamma_\tau)^{-1}(C_s^T \Phi_\tau x(k) + d_0 - s_d(k+1)) \tag{2.44}$$

The control law so designed guarantee that for any $k \geq k^*$, the system states satisfy the inequality

$$|s(k)| = |d(k-1) - d_0| \leq \delta_d. \tag{2.45}$$

Hence, the states of the system settle within a quasi-sliding mode band whose width is less than half the width of the band achieved by the control law proposed in [16]

- Linear Reaching Law: Edwards [10] and also Hui and Zak [18] have given reaching law in another way as

$$s(k+1) = \Phi s(k). \tag{2.46}$$

This reaching law is similar to the Gao's reaching law as well to Sarpturk's reaching law. However, the above reaching law gives an exact description of the desired trajectory toward the sliding surface. Despite the fact that this trajectory cannot be achieved due to the unknown disturbance, the design of the controller is fairly straightforward. Using the system Eq. (2.32) and neglecting the unknown disturbance term, the control law is obtained as

$$u(k) = (\Phi - \Phi_{22})s(k) - \Phi_{21}x_1(k) \tag{2.47}$$

The quasi-sliding mode band is given by

$$\frac{1}{1 - \Phi}(f_{\max}(k)), \tag{2.48}$$

$f_{\max}(k)$ is the maximum value of the disturbance and Φ has all the eigenvalues inside the unit circle.

- Linear Reaching Law with Disturbance Estimation: The smallest quasi-sliding mode band is obtained with the linear reaching law where $\Phi = 0$, but still the quasi-sliding mode band has the same norm as the upper bound for the disturbance. For the Gao's reaching law, the minimum quasi-sliding mode band is even twice the maximum norm of the disturbance. To overcome this problem, a disturbance estimator is introduced. Define disturbance estimator $\tilde{d}(k)$ by:

$$\tilde{d}(k) = \tilde{d}(k - 1) + s(k) - \Phi s(k - 1) \tag{2.49}$$

where $\tilde{d}(k)$ is the estimation of disturbance vector projected on $s(k)$. The above control law (2.47) is changed as

$$u(k) = (\Phi - \Phi_{22})s(k) - \Phi_{21}x_1(k) - \tilde{d}(k) \tag{2.50}$$

In this case, the sliding mode band is given as

$$\frac{1}{1 - \Phi}\delta f(k), \tag{2.51}$$

where $\delta f(k)$ is the maximum rate of change of the disturbance vector.

2.3.2 Output Feedback-Based Discrete-Time Sliding Mode Control

The VSS approach is quite successful in the design of state feedback controller for robust control. But if only the output is accessible, then one needs to utilize output feedback or state estimator (observer). The continuous-time output feedback VSC systems consists of nonlinear and linear parts for the systems with disturbances and/or uncertainties. There have been fundamentally two approaches to design the linear part under the output feedback scheme. The first one is to use state observers [13, 18] and the second one, direct output-based controllers such as static gains types

[6, 17, 18, 26, 27] and dynamic compensators types [11, 12]. Emelyanov et al. [14] proposed an observer to use the very same method as the state feedback VSC. Hui and Zak [18] also constructed an observer-based output feedback controller and even a simpler controller with a static output feedback structure. Kwan [26, 27], Hsu and Lizarralde [17] maintained the linear part as simple as possible, and instead introduced dynamics into the nonlinear part, which allowed them to handle a larger class of matched uncertainties. Edwards and Spurgeon [11], and Edwards et al. [12] considered dynamic variable structure compensators. Especially, they systematically developed a switching surface design method using a dynamic compensator.

The Output Feedback Discrete-time Sliding Mode control (ODSMC) obtained attention recently [5, 32, 33]. Misawa [29] proposed the Observer-Based Sliding Mode Control (OBDSMC) and applied to the position control of single-stage hard disk drive actuators. The algorithm facilitates assignment of eigenvalues for the system matrix which defines the tracking error dynamics inside the boundary layer. Recently, there have been efforts to design the multirate or fast sampled Output Feedback-Based Sliding Mode Control (MROFSMC) where the available output is measured at a faster rate than the input actuation rate and by means of that the states are obtained implicitly [21, 22]. The MROFSMC is applied to various applications like Nuclear Reactor Control, Power Systems, Stepper Motor, Smart Structure, etc.

2.3.3 Multirate Output Feedback Based Sliding Mode Control

In the design of sliding mode controller based on the state feedabck methods, observers are often used to estimate the state vector. The advantage of using an observer is that the observer design can be separated from the controller design and therefore the complete design is simplified. Nonetheless, the introduction of the observer increases the additional complexity. Recently, much work is done on multirate output feedback-based control which guarantee the closed loop stability, while retaining the structural simplicity of the static output feedback [4, 23, 25]. The term multirate includes the situation wherein the system output is sampled at a faster rate compared to the control input.

Consider the continuous-time linear time-invariant system

$$\dot{x}(t) = Ax(t) + Bu(t), \tag{2.52}$$

$$y(t) = Cx(t), \tag{2.53}$$

where $x \in \Re^n$ is the state variable, $A \in \Re^{n \times n}$, $B \in \Re^{n \times m}$ is full rank, $u \in \Re^m$ is the control input, $C \in \Re^{p \times n}$ and $y \in \Re^p$ is the output. We assume that (A, B) is completely controllable and $m < n$. Let, the system in Eqs. (2.52) and (2.53) be discretized at τ sampling instant, then the discrete-time system is given by

$$x(k+1) = \Phi_\tau x(k) + \Gamma_\tau u(k), \tag{2.54}$$

$$y(k) = C_\tau x(k). \tag{2.55}$$

It has been shown by Werner [37] and Janardhanan et al. [21, 22] that the fast sampled output data can be used for state estimation instead of using state observer. In this process, the output measurement is done at N-times faster rate than the input updates and they are related as $\Delta = \tau/N$ where $N \geq \nu$, the observability index of (Φ_τ, C). To realize the fast sampled output system, a fictitious lifted system is constructed for which Δ is considered to be the sampling time at which the output is measured.

Let the system in Eqs. (2.54) and (2.55) be sampled at Δ s is given as

$$x(k+1)\Delta = \Phi_\Delta x(k) + \Gamma_\Delta u(k), \tag{2.56}$$
$$y(k) = Cx(k). \tag{2.57}$$

Definition 2.2 Observability Index: The observability index of a system (Φ, Γ, C) is the smallest positive integer ν such that

$$\text{Rank}\left(\begin{bmatrix} C \\ C\Phi \\ \vdots \\ C\Phi^{\nu-1} \end{bmatrix}\right) = \text{Rank}\left(\begin{bmatrix} C \\ C\Phi \\ \vdots \\ C\Phi^{\nu} \end{bmatrix}\right) \tag{2.58}$$

The relationship between the system parameters of the so-called 'τ' system and the 'Δ' system is given as

$$\Phi_\tau = \Phi_\Delta^N; \qquad \Gamma_\tau = \sum_{i=1}^{N-1} \Phi_\Delta^i \Gamma_\Delta. \tag{2.59}$$

Then, the lifted system with the output sampled at an interval Δ s and the control input update interval τ s would be

$$x(k+1) = \Phi_\tau x(k) + \Gamma_\tau u(k), \tag{2.60}$$
$$y_{k+1} = C_0 x(k) + D_0 u(k), \tag{2.61}$$

where y_k, C_0 and D_0 are defined in [37] as

$$y_k = \begin{bmatrix} y((k-1)\tau) \\ y((k-1)\tau + \Delta) \\ y((k-1)\tau + 2\Delta) \\ \vdots \\ y(k\tau - \Delta) \end{bmatrix}; C_0 = \begin{bmatrix} C \\ C\Phi_\Delta \\ C\Phi_\Delta^2 \\ \vdots \\ C\Phi_\Delta^{N-1} \end{bmatrix}; D_0 = \begin{bmatrix} 0 \\ C\Gamma_\Delta \\ C(\Phi_\Delta\Gamma_\Delta + \Gamma_\Delta) \\ \vdots \\ C\sum_{i=1}^{N-2}\Phi_\Delta^i\Gamma_\Delta \end{bmatrix} \tag{2.62}$$

From Eq. (2.61), we may write

$$x(k) = (C_0^T C_0)^{-1} C_0^T (y_{k+1} - D_0 u(k)).$$

Further,

$$x(k+1) = \Phi_\tau[(C_0^T C_0)^{-1} C_0^T (y_{k+1} - D_0 u(k))] + \Gamma_\tau u(k),$$

$$= \Phi_\tau (C_0^T C_0)^{-1} C_0^T y_{k+1} + \left(\Gamma_\tau - \Phi_\tau (C_0^T C_0)^{-1} C_0^T D_0\right) u(k)$$

$$x(k+1) = L_y y_{k+1} + L_u u(k), \tag{2.63}$$

where

$$L_y = \Phi_\tau (C_0^T C_0)^{-1} C_0^T,$$
$$L_u = \Gamma_\tau - \Phi_\tau (C_0^T C_0)^{-1} C_0^T D_0.$$

Further from Eq. (2.63),

$$x(k) = L_y y_k + L_u u(k-1). \tag{2.64}$$

Thus, the state $x(k)$ can be expressed using fast sampled output stack and past input. The state computation by fast sampled output measurement is better than the conventional discrete-time state observer as it computes the states just in one sampling instant [21] compared to discrete-time observer that takes at least ν instants (ν is the observability index of the system). Moreover, it does not increase the order of the overall system dimension and so reduces the complexity.

Using the Gao's reaching law (2.36), the state feedback control law for the discrete-time LTI system of form (2.54) can be derived as

$$u(k) = -(C_s \Gamma_\tau)^{-1}((C_s \Phi_\tau - C_s + q\tau C_s)x(k) + \rho\tau \operatorname{sgn}(C_s x(k))) \tag{2.65}$$

The above state feedback control algorithm (2.65) can be converted into an output feedback algorithm by the multirate output feedback. Substituting $x(k)$ from (2.64) into (2.65), the multirate output feedback-based sliding mode control law is derived as

$$u(k) = F_y y_k + F_u u(k-1) - (C_s \Gamma_\tau)^{-1} \rho\tau \operatorname{sgn}(C_s L_y y_k + C_s L_u u(k-1)), \tag{2.66}$$

where

$$F_y = -(C_s \Gamma_\tau)^{-1}(C_s \Phi_\tau - C_s + q\tau C_s)L_y,$$
$$F_u = -(C_s \Gamma_\tau)^{-1}(C_s \Phi_\tau - C_s + q\tau C_s)L_u,$$
$$L_y = \Gamma_\tau - L_y D_0,$$
$$L_u = \Phi_\tau (C_0^T C_0)^{-1} C_0^T$$

2.4 Conclusion

In this chapter, first we presented the basic concept of the continuous-time sliding mode and variable structure control. It is evident that a discontinuous control in a continuous time induces a sliding motion in some manifold of the state space. The existence condition of the sliding mode and the stability of the system during sliding mode is also discussed. Moreover, a design method for sliding mode control law based on reaching law is presented. Due to the wide use of digital controllers, it is in demand today to develop discrete-time sliding mode control. In this chapter, we presented the evolution of discrete-time sliding mode along with various reaching laws for the state-based discrete-time sliding mode. The output feedback-based discrete-time sliding mode control strategy is also discussed. Lastly, the multirate output feedback technique for state estimation and the multirate output feedback-based sliding mode control design method are discussed.

References

1. Bartoszewicz A (1995) A comment on A time-varying sliding surface for fast and robust tracking control of second-order uncertain systems. Automatica 31(12):1893–1895
2. Bartoszewicz A (1998) Discrete-time quasi sliding mode control strategies. IEEE Trans Ind Electron 45(04):633–637
3. Bartoszewicz A, Nowacka-Leverton A (2009) Time-varying sliding modes for second and third order systems. Springer, Berlin
4. Chammas AB, Leondes CT (1979) Pole assignment by piecewise constant output feedback. Int J Control 29(01):31–38
5. Chan CY (1999) Discrete adaptive sliding mode control of a class of stochastic systems. Automatica 35:1491–1498
6. Choi HH (2002) Variable structure output feedback control design for a class of uncertain dynamic systems. Automatica 38:335–341
7. Choi SB, Cheong CC, Park DW (1993) Moving switching surfaces for robust control of a second order variable structure systems. Int J Control 58(01):229–245
8. Choi S, Park D, Jayasuriya S (1994) A time-varying sliding surface for fast and robust tracking control of second order uncertain systems. Automatica 30:899–904
9. Drazenovic B (1969) The invariance conditions in variable structure systems. Automatica 05:287–295
10. Edwards C (1998) Sliding mode control: theory and applications. Taylor and Francis, London
11. Edwards C, Spurgeon SK (1998) Compensator based output feedback sliding mode control design. Int J Control 71(04):601–614
12. Edwards C, Spurgeon SK, Hebden RG (2003) On the design of sliding mode output feedback controllers. Int J Control 76(09/10):893–905
13. Emelyanov SV (1967) Variable structure control system (In Russian). Nauka, Moscow
14. Emelyanov SV, Korovin SK, Nersisian AL, Nisenzon YE (1992) Output feedback stabilization of uncertain plants: a variable structure approach. Int J Control 55:61–81
15. Filippov AF, FM A (1988) Differential equations with discontinuous right-hand sides: control systems. Kluwer Academic Publishers, Soviet Union
16. Gao W, Wang Y, Homaifa A (1995) Discrete-time variable structure control system. IEEE Trans Ind Electron 42(02):117–122

17. Hsu L, Lizarralde F (1998) Comments and further results on variable structure output feedback controllers. IEEE Trans Autom Control 43(09):1338–1340
18. Hui S, Zak SH (1999) On discrete-time variable structure sliding mode control. Syst Control Lett 38(4–5):283–288
19. Hung JY, Gao W, Hung JC (1993) Variable structure control: a survey. IEEE Trans Autom Control 40(01):2–22
20. Itkis U (1976) Control systems of variable structure. Wiley, New York
21. Janardhanan S, Bandyopadhyay B (2006) Output feedback sliding mode control for uncertain systems using fast output sampling technique. IEEE Trans Ind Electron 53(05):1677–1682
22. Janardhanan S, Bandyopadhyay B (2007) Multirate output feedback based robust quasi-sliding mode control of discrete-time systems. IEEE Trans Autom Control 52(03):499–503
23. Jury E (1967) A note on multirate sampled-data systems. IRE Trans Autom Control 12(03):319–320
24. Kaynak O, Denker A (1993) Discrete-time sliding mode control in the presence of system uncertainty. Int J Control 57(05):1177–1189
25. Kranc GM (1957) InputOutput analysis of multi-rate feedback system. IRE Trans Autom Control 03:21–28
26. Kwan CM (1996) On variable structure output feedback controllers. IEEE Trans Autom Control 41(11):1691–1693
27. Kwan CM (2001) Further results on variable output feedback controllers. IEEE Trans Autom Control 46(09):1505–1508
28. Milosavljevic C (1985) General conditions for the existence of quasi sliding mode on the switching hyperplane in discrete variable structure systems. Autom Remote Control 46:307–314
29. Richter H, Misawa EA (2002) Boundary layer eigenvalues in observer based discrete-time sliding mode control. In: Proceedings of the American control conference, Anchorage, AK, pp 2985–2936
30. Roy RG, Olgac N (1997) Robust nonlinear control via moving sliding surfaces—n-th order case. In: Proceedings of 36th conference on decision and control, San Diego, pp 943–948
31. Sarpturk SZ, Istefanopulos Y, Kaynak O (1987) On the stability of discrete-time sliding mode control systems. IEEE Trans Control Syst 32(10):930–932
32. Sharav-Schapiro N, Palmor J, Steinberg A (1996) Robust output feedback stabilizing control for discrete uncertain SISO systems. IEEE Trans Autom Control 41(09):1377–1381
33. Sharav-Schapiro N, Palmor J, Steinberg A (1998) Output stabilizing robust control for discrete uncertain systems. Automatica 34(06):731–739
34. Utkin VI (1977) Variable structure systems with sliding mode. IEEE Trans Autom Control 22(02):212–222
35. Utkin VI (1992) Sliding modes in control optimization. Springer, New York
36. Utkin VI, Drakunov SV (1989) On discrete-time sliding modes control. In: Proceedings of IFAC conference on nonlinear control, Capri, Italy, pp 484–489
37. Werner H (1998) Multimodel robust control by fast output sampling—an LMI approach. Automatica 34(12):1625–1630
38. Yilmaz C, Hurmuzlu Y (2000) Eliminating the reaching phase from variable structure control. ASME J Dyn Syst, Meas, Control 122(04):753–757
39. Young KD, Utkin VI, Ozguner U (1999) A control engineers guide to sliding mode control. IEEE Trans Autom Control 07(03):328–342

Chapter 3
Multirate Output Feedback Frequency Shaped SMC: A Switching Type Control Law

Abstract This chapter contains some results from [7, 8] and several additional results on the design method for multirate output feedback-based frequency-shaped sliding mode control. A switching-type reaching law has been used for the vibration suppression of the smart structure. The frequency-shaping approach is used to suppress the vibratory modes of smart structure excited during the sliding mode.

Keywords Frequency-shaped sliding mode control · Multirate output feedback · Switching-type control law · Smart structure · Vibration control

3.1 Brief Review of Discrete-Time Sliding Mode Control Using LQ Approach

Consider the continuous-time linear time-invariant system

$$\dot{x}(t) = Ax(t) + Bu(t), \tag{3.1}$$
$$y(t) = Cx(t), \tag{3.2}$$

where $x \in \Re^n$ is the state variable, $A \in \Re^{n \times n}$, $B \in \Re^{n \times m}$ is full rank, $u \in \Re^m$ is the control input, $C \in \Re^{p \times n}$ and $y \in \Re^p$ are the outputs. We assume that (A, B) is completely controllable and $m < n$.

Let, the system in Eqs. (3.1) and (3.2) be discretized at τ sampling instant then the discrete-time system is given by

$$x(k + 1) = \Phi_\tau x(k) + \Gamma_\tau u(k), \tag{3.3}$$
$$y(k) = C_\tau x(k). \tag{3.4}$$

The system in Eqs. (3.3) and (3.4) is transformed into regular form by $\bar{x} = T_r x$ as:

$$\bar{x}(k + 1) = T_r \Phi_\tau T_r^T \bar{x}(k) + T_r \Gamma_\tau u(k). \tag{3.5}$$

Assuming $\bar{x}^T = [\bar{x}_1^T \ \bar{x}_2^T]$ where $\bar{x}_1 \in \Re^{n-m}$ and $\bar{x}_2 \in \Re^m$ one obtain

$$\bar{x}_1(k + 1) = \Phi_{11} \bar{x}_1(k) + \Phi_{12} \bar{x}_2(k), \tag{3.6}$$

© The Author(s) 2015
A. Mehta and B. Bandyopadhyay, *Frequency-Shaped and Observer-Based Discrete-time Sliding Mode Control*, SpringerBriefs in Applied Sciences and Technology, DOI 10.1007/978-81-322-2238-5_3

$$\bar{x}_2(k+1) = \Phi_{21}\bar{x}_1(k) + \Phi_{22}\bar{x}_2(k) + \Gamma_{\tau 2}u(k), \qquad (3.7)$$

$$y(k) = [C_1 \quad C_2]\bar{x}(k), \qquad (3.8)$$

where

$$T_r\Phi_\tau T_r^T = \begin{bmatrix} \Phi_{11} & \Phi_{12} \\ \Phi_{21} & \Phi_{22} \end{bmatrix},$$

$$T_r\Gamma_\tau = \begin{bmatrix} 0 & \Gamma_{\tau 2} \end{bmatrix}^T,$$

$$C_\tau T_r^T = \begin{bmatrix} C_1 & C_2 \end{bmatrix}.$$

The technique for designing the sliding surface using the linear quadratic (LQ) approach has been given in [10]. Consider a quadratic cost function

$$J_{os} = \frac{1}{2}\sum_{k=0}^{\infty}[\bar{x}_1^T(k)Q_{11}\bar{x}_1(k) + 2\bar{x}_1^T(k)Q_{12}\bar{x}_2(k) + \bar{x}_2^T(k)Q_{22}\bar{x}_2(k)] \qquad (3.9)$$

where Q is a symmetric positive definite matrix

$$T_r Q T_r^T = \begin{bmatrix} Q_{11} & Q_{12} \\ Q_{12}^T & Q_{22} \end{bmatrix}.$$

Without loss of generality, we can assume $Q_{12} = 0$, then

$$J_{os} = \frac{1}{2}\sum_{k=0}^{\infty}[\bar{x}_1^T(k)Q_{11}\bar{x}_1(k) + \bar{x}_2^T(k)Q_{22}\bar{x}_2(k)]. \qquad (3.10)$$

The optimal sliding function can be defined as

$$s_{os} = C_1 K_{os}\bar{x}_1(k) + C_2\bar{x}_2(k), \qquad (3.11)$$

where K_{os} is derived from the optimal solution of the associated Riccati equation.

3.2 State Estimation for LTI System with Uncertainity Using Multirate Output Measurement

Consider the linear time-invariant plant model with matched uncertainity as

$$\dot{x}(t) = Ax(t) + Bu(t) + Ef(t), \qquad (3.12)$$

$$y(t) = Cx(t). \qquad (3.13)$$

The corresponding discrete-time model is given by

$$x(k+1) = \Phi_\tau x(k) + \Gamma_\tau u(k) + E_\tau f(k), \qquad (3.14)$$

$$y(k) = C_\tau x(k), \qquad (3.15)$$

where

$$\Phi_\tau = e^{A\tau}; \quad \Gamma_\tau = \int_0^\tau e^{A(\tau-\theta)} B \, d\theta; \tag{3.16}$$

$$C_\tau = C; \quad E_\tau = \int_0^\tau e^{A(\tau-\theta)} E \, d\theta. \tag{3.17}$$

τ be the sampling time at which the input is updated.

Assumption 1 The triplet $(\Phi_\tau, \Gamma_\tau, C_\tau)$ is controllable and observable. Also the uncertainty is not changing during τ sampling interval.

The output measurement is done at N-times faster rate than the input updates and they are related as $\Delta = \tau/N$ where $N \geq v$, the observability index of (Φ_τ, C). To obtain the fictitious lifted system, let the system in Eqs. (3.12) and (3.13) be sampled at Δ s is given as

$$x(k+1)\Delta = \Phi_\Delta x(k) + \Gamma_\Delta u(k) + E_\Delta f(k),$$
$$y(k) = Cx(k). \tag{3.18}$$

The relationship between the system parameters of the τ sampled system and the Δ sampled system is given as

$$\Phi_\tau = \Phi_\Delta^N; \quad \Gamma_\tau = \sum_{i=1}^{N-1} \Phi_\Delta^i \Gamma_\Delta; \quad E_\Delta = \left(\sum_{i=1}^{N-1} \Phi_\Delta^i\right)^{-1} E_\tau. \tag{3.19}$$

Then the lifted system with the output sampled at an interval Δ s and the control input update interval τ s, would be

$$x(k+1) = \Phi_\tau x(k) + \Gamma_\tau u(k) + E_\tau f(k), \tag{3.20}$$
$$y_{k+1} = C_0 x(k) + D_0 u(k) + C_d f(k), \tag{3.21}$$

where y_k, C_0, D_0 and C_d are defined in [11] as

$$y_k = \begin{bmatrix} y((k-1)\tau) \\ y((k-1)\tau + \Delta) \\ y((k-1)\tau + 2\Delta) \\ \vdots \\ y(k\tau - \Delta) \end{bmatrix}; \quad C_0 = \begin{bmatrix} C \\ C\Phi_\Delta \\ C\Phi_\Delta^2 \\ \vdots \\ C\Phi_\Delta^{N-1} \end{bmatrix}; \tag{3.22}$$

$$D_0 = \begin{bmatrix} 0 \\ C\Gamma_\Delta \\ C(\Phi_\Delta \Gamma_\Delta + \Gamma_\Delta) \\ \vdots \\ C\sum_{i=1}^{N-2} \Phi_\Delta^i \Gamma_\Delta \end{bmatrix}; \quad C_d = \begin{bmatrix} 0 \\ CE_\Delta \\ C(\Phi_\Delta E_\Delta + E_\Delta) \\ \vdots \\ C\sum_{i=1}^{N-2} \Phi_\Delta^i E_\Delta \end{bmatrix} \tag{3.23}$$

From Eq. (3.21), we may write

$$x(k) = (C_0^T C_0)^{-1} C_0^T (y_{k+1} - D_0 u(k) - C_d f(k)).$$

From which

$$x(k + 1) = \Phi_\tau [(C_0^T C_0)^{-1} C_0^T (y_{k+1} - D_0 u(k)) - C_d f(k)] + \Gamma_\tau u(k) + E_\tau f(k),$$

$$= \Phi_\tau (C_0^T C_0)^{-1} C_0^T y_{k+1} + \left(\Gamma_\tau - \Phi_\tau (C_0^T C_0)^{-1} C_0^T D_0 \right) u(k)$$

$$+ \left(E_\tau - \Phi_\tau (C_0^T C_0)^{-1} C_0^T C_d \right) f(k),$$

$$x(k + 1) = L_y y_{k+1} + L_u u(k) + L_d f(k), \tag{3.24}$$

where

$$L_y = \Phi_\tau (C_0^T C_0)^{-1} C_0^T,$$
$$L_u = \Gamma_\tau - \Phi_\tau (C_0^T C_0)^{-1} C_0^T D_0,$$
$$L_d = E_\tau - \Phi_\tau (C_0^T C_0)^{-1} C_0^T C_d.$$

Further from Eq. (3.24),

$$x(k) = L_y y_k + L_u u(k - 1) + L_d f(k - 1).$$

Applying the transformation, $\bar{x} = T_r x$

$$\bar{x}(k) = T_r L_y y_k + T_r L_u u(k - 1) + T_r L_d f(k - 1) \tag{3.25}$$

which can be written in regular form as

$$\bar{x}_1(k) = [I_{n-m} \vdots 0](T_r L_y y_k + T_r L_u u(k - 1) + T_r L_d f(k - 1)), \tag{3.26}$$

$$\bar{x}_2(k) = [0 \vdots I_m](T_r L_y y_k + T_r L_u u(k - 1) + T_r L_d f(k - 1)). \tag{3.27}$$

3.3 Multirate Output Feedback-Based Frequency-Shaped Sliding Mode Control

Consider the plant dynamics (3.20, 3.21) in regular form as

$$\begin{bmatrix} \bar{x}_1(k + 1) \\ \bar{x}_2(k + 1) \end{bmatrix} = \begin{bmatrix} \Phi_{11} & \Phi_{12} \\ \Phi_{21} & \Phi_{22} \end{bmatrix} \begin{bmatrix} \bar{x}_1(k) \\ \bar{x}_2(k) \end{bmatrix} + \begin{bmatrix} 0 \\ \Gamma_{\tau 2} \end{bmatrix} u(k) + \begin{bmatrix} 0 \\ E_{\tau 2} \end{bmatrix} f(k), \tag{3.28}$$

$$y_{k+1} = \bar{C}_0 \bar{x}(k) + \bar{D}_0 u(k) + \bar{C}_d f(k), \tag{3.29}$$

where

$$\bar{C}_0 = C_0 T_r^T, \quad \bar{D}_0 = D_0, \quad \bar{C}_d = C_d.$$

The optimal sliding mode control design based on frequency-shaped cost function and output sampled measurement is achieved by formulating an LQR problem with frequency-dependent weights.

For the cost function in Eq. (3.10), both state variables have fixed weighings for all frequencies. In order to introduce frequency-dependent compensator into sliding mode, the cost function in Eq. (3.10) is rewritten in the frequency domain with frequency-dependent weighing functions as

$$J_{fs} = \frac{1}{2\pi} \int_{-\infty}^{\infty} [\bar{x}_1^*(\omega) Q_{11} \bar{x}_1(\omega) + \bar{x}_2^*(\omega) Q_{22} \bar{x}_2(\omega)] d\omega, \tag{3.30}$$

$$= \frac{1}{2\pi} \int_{-\infty}^{\infty} [(W_1(\omega)\bar{x}_1(\omega))^* W_1(\omega)\bar{x}_1(\omega)$$

$$+ (W_2(\omega)\bar{x}_2(\omega))^* W_2(\omega)\bar{x}_2(\omega)] d\omega, \tag{3.31}$$

$$= \frac{1}{2\pi} \int_{-\infty}^{\infty} [\eta_1^* \eta_1 + \eta_2^* \eta_2] d\omega,$$

where the notation $(\cdot)^*$ denotes a complex conjugate transpose. W_1 is the spectral factor of Q_{11}, i.e., $Q_{ii} = W_i^*(\omega) W_i(\omega)$, $\eta_i(\omega) = W_i(\omega)\bar{x}_i(\omega)$ are the outputs of each filter with transfer functions $W_i(\omega)$ and inputs $\bar{x}_i(\omega)$. In order to obtain a casual optimal solution that minimizes the cost function, both weighing functions $Q_{ii}(\omega)$ are assumed to be proper rational functions of squared frequency ω^2. According to (3.31), if W_i represents a bandpass filter, the minimization of J_{fs} will penalize the component of $\bar{x}_i(\omega)$ in its frequency band [13].

Applying the Parseval's Theorem [9], the time domain counterpart of the cost function (3.30), (3.31) is written as

$$J_{fs} = \frac{1}{2} \sum_{k=0}^{\infty} [\bar{x}_1^*(k) Q_{11} \bar{x}_1(k) + \bar{x}_2^*(k) Q_{22} \bar{x}_2(k)], \tag{3.32}$$

$$= \frac{1}{2} \sum_{k=0}^{\infty} [(W_1(k)\bar{x}_1(k))^* W_1(k)\bar{x}_1(k) + (W_2(k)\bar{x}_2(k))^* W_2(k)\bar{x}_2(k)]. \tag{3.33}$$

It is feasible to perform frequency shaping for $\bar{x}_2(k)$ to achieve required control performance [12]. Hence, $W_1(k)$ can be treated as constant for all frequency and $W_2(k)$ can thus be designed as a required compensator expressed as

$$x_\omega(k+1) = A_\omega x_\omega(k) + B_\omega \bar{x}_2(k), \tag{3.34}$$

$$= A_\omega x_\omega(k) + B_\omega[0 \vdots I_m](T_r L_y y_k + T_r L_u u(k-1) + T_r L_d f(k-1)),$$

$$\eta_2 = C_\omega x_\omega(k) + D_\omega \bar{x}_2(k), \tag{3.35}$$

where $A_\omega \in \mathfrak{R}^{q \times q}$, $B_\omega \in \mathfrak{R}^{q \times m}$, $C_\omega \in \mathfrak{R}^{l \times q}$ and $D_\omega \in \mathfrak{R}^{l \times m}$ are to be selected.

With $\bar{x}_2(k)$ as input and η_2 as output, we may write above filter dynamics in transfer function form as

$$R(z) = C_\omega(zI - A_\omega)^{-1}B_\omega + D_\omega. \tag{3.36}$$

Consider an augmented system formed from Eqs. (3.28) and (3.34) as

$$x_e(k + 1) = A_e x_e(k) + B_e \bar{x}_2(k) \tag{3.37}$$

where,

$$x_e(k) = \begin{bmatrix} x_w(k) \\ \bar{x}_1(k) \end{bmatrix}; \quad A_e = \begin{bmatrix} A_\omega & 0 \\ 0 & \Phi_{11} \end{bmatrix}; \quad B_e = \begin{bmatrix} B_\omega \\ \Phi_{12} \end{bmatrix}. \tag{3.38}$$

Define a new hyperplane in augmented state space, formed from the plant and compensator state space as

$$s(k) = Kx_e(k) + \bar{x}_2(k), \tag{3.39}$$
$$= K_\omega x_\omega + K_1 \bar{x}_1(k) + \bar{x}_2(k),$$
$$= K_\omega x_\omega(k) + K_x \bar{x}(k), \tag{3.40}$$

The sliding surface may be represented in terms of fast sampled output and input as

$$s(k) = K_\omega x_\omega(k) + K_x(T_r L_y y_k + T_r L_u u(k-1) + T_r L_d f(k-1)) \tag{3.41}$$

where, $K_x = [K_1 \quad I_m]$.

The sliding gain K in (3.39) may be designed either by pole placement technique or by linear quadratic (LQ) approach.

For using the LQ approach, define the cost function for the augmented system dynamics as

$$J_{fs} = \frac{1}{2}\sum_{k=0}^{\infty}[x_e^T(k)Q_e x_e(k) + 2x_e^T(k)N_e \bar{x}_2(k) + \bar{x}_2^T(k)R_e \bar{x}_2(k)], \tag{3.42}$$

where

$$Q_e = \begin{bmatrix} C_\omega^T C_\omega & 0 \\ 0 & Q_{11} \end{bmatrix}; \quad N_e = \begin{bmatrix} C_\omega^T D_\omega \\ 0 \end{bmatrix}; \quad R_e = D_\omega^T D_\omega. \tag{3.43}$$

The solution of the above equation is obtained as

$$K = (R_e + B_e^T P_e B_e)^{-1}(B_e^T P_e A_e + N_e^T)$$

derived from the associated Riccati equation.

The dynamics of the sliding surface which incorporates the compensator dynamics is obtained from Eq. (3.39) for $s(k) = 0$,

$$\bar{x}_2(k) = -K x_e(k). \tag{3.44}$$

Substituting in Eq. (3.37)

$$x_e(k + 1) = (A_e - B_e K) x_e(k). \tag{3.45}$$

The locations of the eigen values of the $A_e - B_e K$ decides the behavior of the augmented system on the sliding surface. Also from the above, it is clear that the introduction of compensator has produced more design freedom. After designing the sliding surface with associated compensator, the control law must be designed to generate sliding motion.

As discussed in Sect. 2.4.1, there are two schools of thought for discrete-time sliding mode control. 1. Switching type and 2. Nonswitching type. In switching type of law (Gao's law), the system trajectory crosses the switching surface in finite sampling instant and once it crosses, it recrosses the surface in the next sampling instant and thereby it approaches the origin in a zigzag motion about the sliding surface, and remains in a band. This kind of motion exists for system without or with matched uncertainity. Whereas in nonswitching type of law (Golo's and Milosavljevic's law, Bartolini, Ferrara, and Utkin's law, and Bartoszewicz's law), the system trajectory for system without uncertainity exactly reaches the switching surface in finite sampling instant and thereafter it slides along the surface. For uncertain system, it remains in a boundary layer of the sliding surface. The multirate output feedback-based frequency-shaped sliding mode control law can be designed by any one of the switching or nonswitching type of reaching laws. In this chapter, we used the switching type reaching law for designing FSSMC for the vibration suppression of smart structure.

3.4 The Controller Design Using Switching Type Control Law

3.4.1 Brief Review of Switching Type Control Law

In order to design a DSMC, the Gao et al. [3] has proposed switching type reaching law as

$$s(k + 1) = (1 - q\tau)s(k) - \rho\tau\text{sgn}(s(k)), \tag{3.46}$$

where, $\tau > 0$ is the sampling time, $q > 0$, $\rho > 0$, and $1 - q\tau > 0$.

The DSMC is required to achieve the following performances [3]:

1. Starting from any initial state, the trajectory will move monotonically toward the switching plane and cross it in finite time.

2. Once the trajectory has crossed the switching plane, it will cross the plane again in every successive sampling period, resulting in a zigzag motion about the switching plane.
3. The size of each successive zigzag step is nonincreasing and the trajectory stays within a specified band.

With the above reaching law (3.46) and the sliding surface $s(k) = C_s x(k)$, the control law is derived as

$$u(k) = -(C_s \Gamma_\tau)^{-1}[C_s \Phi_\tau x(k) - (1 - q\tau)s(k) + \rho\tau \text{sgn}(s(k))]. \qquad (3.47)$$

It is possible to derive quasi-sliding mode band (QSMB) for $s(k)$ that achieves the above DSMC performance (2) as

$$\delta_s = \frac{\rho\tau}{2 - q\tau}. \qquad (3.48)$$

3.4.2 The FSSMC Design Using Switching Type Reaching Law

In order to design a FSSMC, the modified Gao's reaching law [5] approach is adopted as

$$s(k+1) - s(k) = -q\tau s(k) - \rho\tau \text{sgn}(s(k)) + \tilde{f}(k) - f_0 - f_1 \text{sgn}(s(k)), \quad (3.49)$$

where,

$$f_l \leq \tilde{f}(k) = K_x T_r E_\tau f(k) + K_w B_w E_{\tau 2} f(k-1) \leq f_u,$$

and also the mean and spread are given as

$$f_0 = \frac{f_l + f_u}{2},$$

$$f_1 = \frac{f_u - f_l}{2}.$$

From the above reaching law (3.49) and augmented sliding surface (3.40), the dynamic output feedback control law is derived as,

$$\begin{aligned}
u(k) = - (K_x \Gamma_\tau)^{-1}\{(K_\omega A_\omega + (1 - q\tau)K_\omega)x_\omega(k) \\
+ (K_x \Phi_{\text{reg}} + (1 - q\tau)K_x)\bar{x}(k) + (K_\omega B_\omega)\bar{x}_2(k) \\
+ f_0 + (f_1 + \rho\tau)\text{sgn}(s(k))\},
\end{aligned} \qquad (3.50)$$

where Φ_{reg} is the regular form of matrix Φ_τ.

The above state feedback-based control law (3.50) can be converted to output feedback-based control law by substituting for $\bar{x}(k)$ from Eqs. (3.26) and (3.27).

However, due to the presence of the uncertainty terms $T_r L_d f(k)$ and $[0 \quad I_m]T_r L_d f(k)$, the control law would not be implementable. Hence, for achieving a multirate output feedback-based control algorithm, the control law has to be redesigned from a modified reaching law suited for output feedback.

Consider the modified reaching law as

$$
\begin{aligned}
s(k+1) - s(k) = &-q\tau s(k) - \rho\tau\mathrm{sgn}(s(k)) + \tilde{g}(k-1) \\
&+ \tilde{f}(k) - f_0 - g_0 - (g_1 + f_1)\mathrm{sgn}(s(k)),
\end{aligned} \tag{3.51}
$$

where,

$$
\begin{aligned}
g_l \le \tilde{g}(k) = &(K_x\Phi_{\text{reg}} + (1 - q\tau)K_x)T_r L_d f(k) \\
&+ (K_\omega B_\omega)[0 \quad I_m]T_r L_d f(k) \le g_u,
\end{aligned} \tag{3.52}
$$

where,

$$
g_0 = \frac{g_l + g_u}{2}
$$

$$
g_1 = \frac{g_u - g_l}{2}
$$

The control law may be formulated from (3.51) as

$$
\begin{aligned}
u(k) = &-(K_x\Gamma_\tau)^{-1}\{(K_\omega A_\omega + (1 - q\tau)K_\omega)x_\omega(k) \\
&+ (K_x\Phi_{\text{reg}} + (1 - q\tau)K_x)\bar{x}(k) + (K_\omega B_\omega)\bar{x}_2(k) \\
&- \tilde{g}(k-1) + g_0 + f_0 + (g_1 + f_1 + \rho\tau)\mathrm{sgn}(s(k))\}.
\end{aligned} \tag{3.53}
$$

The uncertainty term $\tilde{g}(k-1)$ will get canceled in the above control law (3.53) while replacing state with the multirate output Eq. (3.25). But still for implementation of the control law, we need to calculate $s(k)$ that contains the uncertainty terms $T_r L_d f(k)$ and $[0 \quad I_m]T_r L_d f(k)$. The obtained value of $\mathrm{sgn}(s(k))$ may be misleading so the switching function $\mathrm{sgn}(s(k))$ is replaced with $\mathrm{sgn}(\bar{s}(k))$ which can be computed from

$$
\begin{aligned}
\bar{s}(k) &= K_\omega x_\omega(k) + K_x(T_r L_y y_k + T_r L_u u(k-1)) + l_0 \\
&= s_k - l(k-1) + l_0
\end{aligned} \tag{3.54}
$$

where,

$$
l_l \le K_x T_r L_d f(k-1) + K_w B_w[0 \quad I_m]T_r L_d f(k-2) = l(k-1) \le l_u, \tag{3.55}
$$

$$
l_0 = \frac{l_l + l_u}{2},
$$

$$
l_1 = \frac{l_u - l_l}{2},
$$

The value of $\text{sgn}(s(k)) = \text{sgn}(\bar{s}(k))$ whenever $|\bar{s}(k)| > l_1$ but when the value of $|\bar{s}(k)| < l_1$ the sign of $(s(k))$ cannot be determined accurately as $s(k)$ is in the range of $s(k) \pm l_1$. Hence, in order to assure quasi-sliding mode band in spite of this uncertainity, the width of the quasi-sliding mode band should be such that it encompasses this ambiguous band of $|s(k)| \le 2l_1$ which gives the addition constrain on the controller parameters.

The modified control law using (3.54) can be written as

$$
\begin{aligned}
u(k) = - (K_x \Gamma_\tau)^{-1} \{ & (K_\omega A_\omega + (1 - q\tau) K_\omega) x_\omega(k) \\
& + (K_x \Phi_{\text{reg}} + (1 - q\tau) K_x) \bar{x}(k) + (K_\omega B_\omega) \bar{x}_2(k) \\
& - \tilde{g}(k-1) + g_0 + f_0 + (g_1 + f_1 + \rho\tau) \text{sgn}(\bar{s}(k)) \}. \quad (3.56)
\end{aligned}
$$

The above control law (3.56) is obtained in terms of output feedback by replacing the terms $\bar{x}(k)$ and $\bar{x}_2(k)$ using Eqs. (3.25) and (3.27) as

$$
\begin{aligned}
u(k) = - (K_x \Gamma_\tau)^{-1} \{ & (K_\omega A_\omega + (1 - q\tau) K_\omega) x_\omega(k) \\
& + (K_x \Phi_{\text{reg}} + (1 - q\tau) K_x)(T_r L_y y_k + T_r L_u u(k-1)) \\
& + (K_\omega B_\omega)([0 \quad I_m](T_r L_y y_k + T_r L_u u(k-1))) \\
& + g_0 + f_0 + (g_1 + f_1 + \rho\tau) \text{sgn}(\bar{s}(k)) \}. \quad (3.57)
\end{aligned}
$$

So the control law is free from any uncertainty term. The quasi-sliding mode band may be obtained as

$$
\delta_y < \frac{\rho\tau + l_1 + 2(f_1 + g_1)}{1 - q\tau}. \quad (3.58)
$$

Thus, the multirate output feedback-based control algorithm has an increased bound on the QSM band in comparison to the state feedback-based QSM band (3.48). Moreover, it is observed that the QSM band $\delta_y > l_1$, due to the fact that $(1 - q\tau) < 1$. Thus, even with the uncertainty of the sign of $s(k)$ in the range $|s(k)| < l_1$, the system still remains in the QSM band δ_y, even when $s(k)$ is replaced by $\bar{s}(k)$ in the signum function present in the control algorithm (3.57).

3.5 State Space Model of the Smart Structure

Consider a flexible cantilever beam embedded with piezoelectric sensor/actuator along the length of the beam as shown in Fig. 3.1. The optimum location of the piezoelectric sensor/actuator may be obtained from [6]. Manjunath [6] observed that modeling a smart structure by including the sensor/actuator mass and stiffness and by varying its location on the beam from the free end to the fixed end introduces a considerable change in the system's structural vibration characteristics. Further,

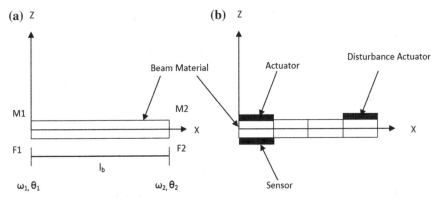

Fig. 3.1 a Cantilever beam, b smart beam

from the responses of the various models of each system, it is observed that when the piezoelectric element is placed near the clamped end, i.e., the fixed end, the sensor output voltage is greater. This is due to the heavy distribution of the bending moment near the fixed end for the fundamental mode, thus leading to a larger strain rate. The sensor voltage is very less when the sensor/actuator pair is located at the free end. The sensitivity of the sensor/actuator pair depends on its location on the beam.

The dimensions of flexible beam and sensor/actuator are given in Tables 3.1 and 3.2, respectively [2, 6]. The dynamic equation of the smart structure is obtained by using both regular beam element and piezoelectric beam elements with following assumptions. The mass and stiffness matrices of the smart structure includes sensor/

Table 3.1 Characteristics of flexible beam

Length (m)	T_b	0.40
Width (m)	B	0.0135
Thickness (mm)	t_b	0.001
Young's modulus (GPa)	E_b	71
Density (Kg/m^2)	ρ_b	2,700
First natural frequency (Hz)	f_1	5.04
Second natural frequency (Hz)	f_2	32.84

Table 3.2 Properties of piezo sensor/actuator

Length (m)	l_b	0.0765
Width (m)	B	0.0135
Thickness (mm)	t_p	0.0005
Young's modulus (GPa)	E_p	47.62
Density (Kg/m^3)	ρ_p	7,500
Piezoelectric strain constant (m V^{-1})	d_{31}	-247×10^{-12}
Piezoelectric stress constant (V m N^{-1})	g_{31}	-9×10^{-3}

actuator mass and stiffness, bonding sensor/actuator layer, and master structure. The cable capacitance between the sensor and the signal conditioning device is considered negligible. Also the temperature effect is neglected. The signal conditioning device gain (Gc) is assumed as 100.

The entire structure is modeled in state space form using the Finite Element Method (FEM) [4] by dividing the structure into four finite elements. The dynamic equation of motion of smart structure obtained using FEM is given by

$$M\ddot{\varphi} + C\dot{\varphi} + K\varphi = f_{\text{ext}} + f_{\text{ctr}} = f^t, \tag{3.59}$$

where M, C, K, φ are the mass, damping, stiffness matrices of smart structure and vector of displacement, respectively. The generalized coordinates are introduced into Eq. (3.59) using a transformation $\varphi = Tg$ in order to reduce it further such that the equation represents the dynamics of two vibratory modes of the flexible cantilever beam. The nodal displacement vector in Eq. (3.59) is given by

$$\varphi = [\omega_1 \quad \theta_1 \quad \omega_2 \quad \theta_2]^T \tag{3.60}$$

where ω_1, θ_1, and ω_2, θ_2 are the DOFs at the fixed end and free end, respectively. The dynamic equation of the smart cantilever beam is developed as

$$M^*\ddot{g} + C^*\dot{g} + K^*g = f_{\text{ext}}^* + f_{\text{ctr}}^* \tag{3.61}$$

The sensor equation is modeled as $V^s(t) = P^T q = y(t)$ where P^T is a constant vector, which depends on the piezoelectric sensor characteristics and on the position of the sensor location on the beam. Using the state space techniques, the SISO state space model is obtained as

$$\dot{x}(t) = Ax(t) + Bu(t) + Er(t), \tag{3.62}$$
$$y(t) = Cx(t) + Du(t), \tag{3.63}$$

with

$$A = \begin{bmatrix} 0 & I \\ -M^{*-1}K^* & -M^{*-1}C^* \end{bmatrix}_{(4\times4)}; B = \begin{bmatrix} 0 \\ M^{*-1}T^Th \end{bmatrix}_{(4\times n)};$$

$$C = \begin{bmatrix} 0 & P^TT \end{bmatrix}_{(n\times4)}; D = \text{A Null matrix}; E = \begin{bmatrix} 0 \\ M^{*-1}T^Tf \end{bmatrix}_{(4\times n)}.$$

where the various parameters of the smart beam M^*, K^*, C^*, f_{ext}^*, f_{ctr}^*, f^t, A, B, C, $x(t)$, $y(t)$, T, P^T, and h^T given in (3.59) and (3.61) represent generalized mass matrix, the generalized stiffness matrix, the generalized structural model damping matrix, the generalized external force, the generalized control force, the total force vector, system matrix, input matrix, output matrix, state vector, output vector, model matrix containing the eigen vector, and the constant vector. n is the number of sensor/actuator pair.

In spite of extensive studies and achievements, the analytical model-based approaches have been highly doubtful under high precision requirements because of the difficulty in simulating the properties of these complicated systems. The finite element model-based methods are usually time-consuming and their applications for accurate control are sometimes hindered by factors such as the assumption of perfect bonding at the interface between the structure and transducers. In most cases, these traditional modeling approaches are intractable and even impossible for highly complex structures. System identification is an established modeling tool in engineering and numerous successful applications have been reported [1]. Smart structures represent an interesting challenge for system identification methods based on recursive least squares estimation to identify the dynamics of smart structures.

The unknown parameters of the smart structure dynamics are estimated using an online identification method, which is proven to be more universal and feasible than analytical and numerical models for the present system. The recursive least squares (RLS) method based on the ARX model is used for linear system identification, which is easy to implement and has fast parameter convergence. The continuous state space model [2] derived from the identified fourth-order ARX model parameter is

$$
A = \begin{bmatrix} 76.9893 & 71.5731 & -45.5632 & 71.9048 \\ -136.1042 & 6.1271 & 116.6837 & -116.7537 \\ 115.7932 & -116.2021 & -6.5425 & 136.6781 \\ -70.8876 & 45.1268 & -71.2161 & -77.5364 \end{bmatrix}; \quad B = \begin{bmatrix} 0.2046 \\ 0.1955 \\ -0.4427 \\ -0.0299 \end{bmatrix};
$$
(3.64)

$$
C = \begin{bmatrix} 1 & 0 & 0 & 0 \end{bmatrix}; \quad D = 0; \quad E = \begin{bmatrix} 0.0029 \\ 0.0265 \\ -0.0664 \\ 0.0588 \end{bmatrix}
$$
(3.65)

3.6 Controller Design Procedure

The frequency-shaped sliding mode control with fast sampled output measurement is designed to reduce the amplitude of vibration of the cantilever beam at resonance frequencies. The discrete-time system is obtained by sampling the system in Eqs. (3.62) and (3.63) at sampling rate 0.01 s. The sensor output is sampled N times faster at 0.0025 s as $(\Delta = \tau/N)$ where $N = 4$. As shown in the frequency response of the smart structure model (Fig. 3.2), there are two modes of natural frequencies at 5.04 Hz (31.9 rad/s) and 32.84 Hz (204 rad/s), respectively. The frequency content

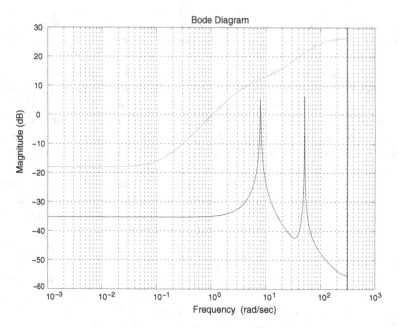

Fig. 3.2 Frequency response of the smart structure model and compensator

of the optimal sliding mode is shaped by using a frequency-dependent compensator presented in Eq. (3.34), such that a higher gain can be obtained at the natural frequencies. The compensator is designed as

$$A_\omega = \begin{bmatrix} 0.8 & 0.4 \\ 0 & 0.3169 \end{bmatrix}; \qquad B_\omega = \begin{bmatrix} 0.05212 \\ 0.03112 \end{bmatrix}$$

$$C_\omega = \begin{bmatrix} 0.05212 & -0.03112 \end{bmatrix}; \qquad D_\omega = 0.003125 \qquad (3.66)$$

With the above compensator, the augmented system (3.37) is constructed and by solving the Riccati equation, the gain K is obtained as

$$K = [0.0740 \ 0.1759 \ 3.1417 \ 0.8021 \ 2.2340]$$

The eigen values of the $(A_e - B_e K)$ are $[-0.0001 \ 0.6075 \ 0.7873 \pm 0.1049i \ 0.9473]$. Thus the augmented system in the sliding mode is stable.

Using the above gain K and considering the values of q and ε as 15 and 0.05, respectively, the control law (3.57) is designed as

$$u(k) = -(0.0047)^{-1}\{[0.1147 \ 0.2671]x_\omega(k) + [5.1077 \ 1.4505 \ 2.9823 \ 2.6925]$$

$$(T_r L_y y_k + T_r L_u u(k-1)) + (0.0648)([0 \ I_m]$$

$$(T_r L_y y_k + T_r L_u u(k-1))) + g_0 + f_0 +$$

$$(g_1 + f_1 + \varepsilon\tau)\text{sgn}(\bar{s}(k))\}. \qquad (3.67)$$

Fig. 3.3 Experimental setup for the smart structure

where,

$$
L_y = \begin{bmatrix} -0.9976 & 3.7322 & -5.4729 & 3.7367 \\ -21.5126 & 65.5140 & -70.4864 & 26.4146 \\ -21.4627 & 53.9758 & -47.4234 & 14.9526 \\ -0.9904 & 0.0000 & -0.0000 & 0.0000 \end{bmatrix} ; L_u = \begin{bmatrix} -0.0000 \\ -0.0039 \\ -0.0024 \\ -0.0005 \end{bmatrix}
$$

3.7 Controller Implementation

The experimental setup is shown in the Fig. 3.3. Two piezoceramic patches are surface bonded at a distance of 5 mm from the fixed end of the beam. The patch bonded on the bottom surface acts as a sensor and the one on the top surface acts as an actuator. An excitation input is applied to the structure through another piezoceramic patch which is bonded on the top surface at a distance of 370 mm from the fixed end. The dimensions and properties of the beam and piezoceramic patches are given in Tables 3.1 and 3.2. A piezoceramic of type SP-5H which is equivalent to NAVY TYPE VI, from Sparkler Ceramics Pvt. Ltd, India, is used in the experimental setup. The sensor output signal which is conditioned using a piezo-sensing system is given as analog input to the dSPACE1104 controller board. The control algorithm is developed using simulink software and implemented in real time

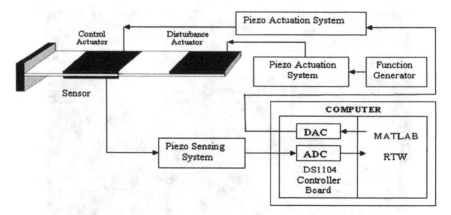

Fig. 3.4 Block diagram of the control algorithm implementation

on dSPACE1104 using RTW and dSPACE real-time interface tools (Fig. 3.4). The simulink software is used to build control block diagrams and real-time workshop is used to generate C code from the simulink model. The C code is then converted to target specific code by the real-time interface and target language compiler supported by dSPACE1104. This code is then deployed on a rapid prototype hardware system, to run the hardware in the loop simulation. The control signal generated from simulink is interfaced to a piezo-actuation system through the configurable analog input/output unit of dSPACE1104. The excitation signal is applied from the simulink environment through a DAC port of the dSPACE controller board.

3.8 Simulation and Experimental Results

To show the performance of the proposed controller, the beam is excited by the excitation frequency. The excitation frequency is also changed from the first mode to second mode after 11 s approximately. The simulation results and the experimental results with switching type FSSMC are shown in the Figs. 3.5–3.7. The results show that the magnitude of vibration reduces to ± 0.08 and ± 0.2 V during the second vibratory mode excitement compared to ± 0.2 and ± 1.2 V in open loop, respectively. Furthermore to illustrate the effectiveness of the controller the frequency response of the system is studied, when the beam is excited with a signal having first and second mode of frequencies (Fig. 3.8). The response shows vibration reduction of approximately 5 dB at the first natural frequency and 11 dB at the second natural frequency.

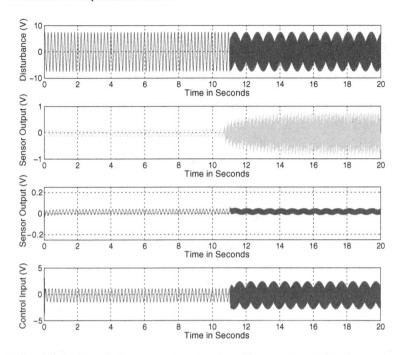

Fig. 3.5 Simulation results with switching type multirate output feedback FSSMC when the excitation frequency is changed from first mode to second mode after 10 s approximately

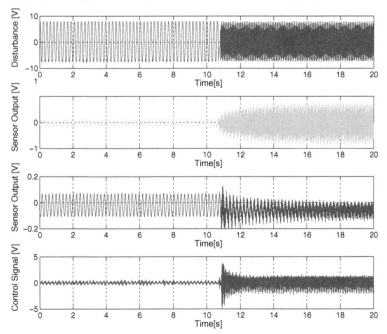

Fig. 3.6 Experimental results with switching type multirate output feedback FSSMC when the excitation frequency is changed from first mode to second mode after 10 s approximately

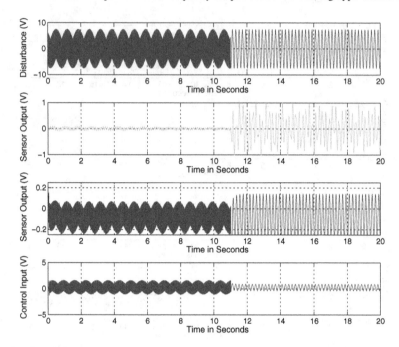

Fig. 3.7 Simulation results with switching type multirate output feedback FSSMC when the excitation frequency is changed from second mode to first mode after 10 s approximately

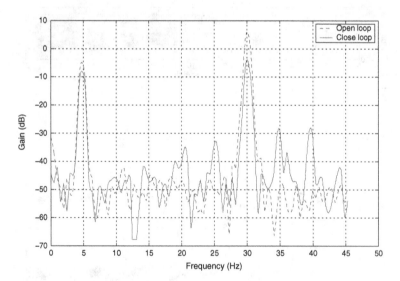

Fig. 3.8 Experimental frequency response

3.9 Conclusion

The chapter proposed a new switching function-based method of designing the multirate output feedback sliding mode control law when the LQ weighings are not constant for all frequencies. The proposed controller uses present output and past input for observing the states which makes it suitable for the practical implementation of the algorithm. Also it does not include any observer dynamics in the closed loop. Another advantage of these controllers is that it has one more choice of tuning the performance of the system by selecting the compensator dynamics introduced in the sliding surface which makes it linear dynamical surface.

References

1. Bu X, Ye L, Su Z, Wang C (2003) Active control of a flexible smart beam using a system identification technique based on ARMAX. Smart Mater Struct 12:845–850
2. Ezhilarasi D, Umapathy M, Bandyopadhyay B (2006) Design and experimental evaluation of piecewise output feedback control for structural vibration suppression. Smart Mater Struct 15(06):1927–1938
3. Gao W, Wang Y, Homaifa A (1995) Discrete-time variable structure control system. IEEE Trans Ind Electron 42(02):117–122
4. Hwang WS, Park HC (1993) Finite element modelling of piezoelectric sensors and actuators. AIAA J 31(05):930–937
5. Janardhanan S, Bandyopadhyay B (2007) Multirate output feedback based robust quasi-sliding mode control of discrete-time systems. IEEE Trans Autom Control 52(03):499–503
6. Manjunath TC (2006) Thesis on multirate output feedback control of cantilever beams using smart structure concept. PhD thesis, Systems and Control Engineering, IIT, Mumbai
7. Mehta A, Bandyopadhyay B (2006) Multirate output feedback based frequency shaped sliding mode control. In: Proceedings of IEEE international conference on industrial technology (ICIT2006), Mumbai, India, pp 2658–2662
8. Mehta A, Bandyopadhyay B (2010) The design and implementation of output feedback based frequency shaped sliding mode controller for the smart structure. In: Proceedings of 2010 IEEE symposium on industrial electronics (ISIE2010), Bari, Italy, pp 353–358
9. Oppenheim AV, Willsky AS, Young IT (1983) Multirate output feedback based robust quasi sliding mode control of discrete-time systems. Prentice Hall, Englewood Cliffs
10. Utkin VI (1992) Sliding modes in control optimization. Springer, New York
11. Werner H (1998) Multimodel robust control by fast output sampling—an LMI approach. Automatica 34(12):1625–1630
12. Wu WC, Liu TS (2005) Frequency shaped sliding mode control for flying height of pickup head in near field optical disc drives. IEEE Trans Magn 41(02):1061–1063
13. Young KD, Ozguner U (1993) Frequency shaping compensator design for sliding mode. Int J Control 57(05):1005–1019

Chapter 4
Multirate Output Feedback Frequency-Shaped SMC: A Nonswitching Type Control Law

Abstract This chapter contains the design method for multirate output feedback frequency-shaped sliding mode control-based on nonswitching type reaching law. The controller is designed for the vibration suppression of the smart structure. The results show that the algorithm damp out the vibrations effectively with the proposed algorithm.

Keywords Frequency-shaped sliding mode control · Multirate output feedback · Nonswitching type control law · Smart structure · Vibration control

4.1 Introduction

An important property of discrete-time systems is that the control signal is computed and applied only at the sampling instants. However, when a switching-based control is used which can change only at the sampling instant, the same will not be able to bring a motion along the sliding surface. This can bring a motion about the sliding surface in a zigzag way which is termed as quasi-sliding mode. It has been shown in [1, 3] that using equivalent control, an exact sliding motion can be obtained, i.e., the motion will slide along the sliding surface when there is no disturbance in the system. For the system affected by a bounded disturbance, motion will remain within a band of the sliding surface. The nonswitching type sliding mode control law or equivalent control law can be obtained from the following reaching law:

$$s(k+i) = 0 \quad \text{for some} \quad i \geq 1, \tag{4.1}$$

for $s(k) = C_s x(k)$.

The resultant control law, in the case of $i = 1$, would be of the form

$$u(k) = -(C_s^T \Gamma_\tau)^{-1}(C_s^T \Phi_\tau x(k)) \tag{4.2}$$

It seems that the control law is not of variable structure type but can bring sliding motion for discrete-time system. Hence it can be seen that though the sliding mode control began with variable structure control, during the course of its development, the concept of sliding mode for discrete-time system became independent of VSC.

A. Mehta and B. Bandyopadhyay, *Frequency-Shaped and Observer-Based Discrete-time Sliding Mode Control*, SpringerBriefs in Applied Sciences and Technology, DOI 10.1007/978-81-322-2238-5_4

This chapter contains some results from [4, 5] and a few additional results on the design method for multirate output feedback frequency-shaped sliding mode control based on nonswitching-type reaching law for the vibration suppression of the smart structure.

4.2 The Controller Design Using Nonswitching Type Control Law

4.2.1 Brief Review of Nonswitching Type Control Law

Bartoszewicz [1] proposed a nonswitching type discrete-time sliding mode control strategy that guarantees finite time convergence of the state trajectory to the sliding surface. Consider a discrete-time system

$$x(k+1) = \Phi_\tau x(k) + \Gamma_\tau u(k) + E_\tau f(k) \tag{4.3}$$
$$y(k) = Cx(k) \tag{4.4}$$

Assumption 2 The triplet $(\Phi_\tau, \Gamma_\tau, C)$ is controllable and observable. Also $f(k < 0) = 0$.

Bartoszewicz proposed a nonswitching type reaching law for a system with uncertainity as

$$s(k+1) = \tilde{f}(k) - f_0 + s_d(k+1), \tag{4.5}$$

where the uncertainty term $\tilde{f}(k) = E_\tau f(k)$ is defined as $f_l \leq \tilde{f}(k) \leq f_u$ with f_l as lower bound and f_u as upper bound and also the mean and spread are given as

$$f_0 = \frac{f_l + f_u}{2},$$

$$\delta_f = \frac{f_u - f_l}{2}.$$

$s_d(k)$ is a priori known function such that the following applies:

- If $s(0) > 2\delta_f$ then

$$s_d(0) = s(0) \tag{4.6}$$
$$s_d(k) \cdot s_d(0) \geq 0 \quad \text{for any} \quad k \geq 0 \tag{4.7}$$
$$s_d(k) = 0 \quad \text{for any} \quad k \geq k^* \tag{4.8}$$
$$|s_d(k+1)| < |s_d(k)| - 2\delta_f \quad \text{for any} \quad k \leq k^* \tag{4.9}$$

The above relations state that the time-dependent hyperplane $s_d(k)$ monotonically, and in finite time converges from its initial position $s_d(0)$ to the sliding surface

$s(k) = 0$. Furthermore, in each control step, the hyperplane moves by the distance greater than $2\delta_d$. This implies that the reaching condition is satisfied, even in the case of the worst combination of disturbance.

• Otherwise, i.e., if $s(0) < 2\delta_f$ then $s_d(k) = 0$ for any $k \geq 0$.

The constant k^* in the above relations is a positive integer chosen by the designer in order to achieve a good trade-off between the fast convergence rate of the system and the magnitude of the control required to achieve this convergence rate.

The control law that satisfies the reaching law in Eq. (4.5) can be computed using Eq. (4.3) as

$$u(k) = -(C^T \Gamma_\tau)^{-1}(C^T \Phi_\tau x(k) + f_0 - s_d(k+1)) \tag{4.10}$$

The control law is designed to guarantee that for any $k \geq k^*$, the system states satisfy the inequality

$$|s(k)| = |f(k-1) - f_0| \leq \delta_d \tag{4.11}$$

Hence, the states of the system settle within a quasi-sliding mode band whose width is less than half the width of the band achieved by control law proposed in [2].

4.2.2 The FSSMC Design Using Nonswitching Type Control Law

In order to design FSSMC using nonswitching type of control law, the Bartoszevicz's reaching law approach is adopted as

$$s(k+1) = \tilde{f}(k) - f_0 + s_d(k+1), \tag{4.12}$$

where $s_d(k)$ is an a priori known function as defined in Sect. 4.2.1.

Consider the system described by Eqs. (3.28 and 3.29) and define the unknown term $\tilde{f}(k)$ as

$$f_l \leq \tilde{f}(k) = K_w B_w [0 \quad I_m] T_r L_d f(k-1) + K_x \Phi_{reg} T_r L_d f(k-1)$$
$$+ K_x T_r E_\tau f(k) \leq f_u, \tag{4.13}$$

and also the mean and spread are given as

$$f_0 = \frac{f_l + f_u}{2},$$

$$f_1 = \frac{f_u - f_l}{2}.$$

From the above reaching law (4.12) and augmented sliding surface (3.40), the state feedback control law is derived as

$$u(k) = -(K_x \Gamma_{reg})^{-1}\{(K_\omega A_\omega x_\omega(k) + K_\omega B_w \bar{x}_2(k)$$

$$+ K_x \Phi_{reg} \bar{x}_2(k) + f_0 - s_d(k+1)\}, \qquad (4.14)$$

The output feedback control law is obtained by replacing the states with multirate output relationships as

$$u(k) = -(K_x \Gamma_{reg})^{-1}\{(K_\omega A_\omega x_\omega(k) + K_\omega B_w[0 \quad I_m](T_r L_y y_k + T_r L_u u(k-1))$$

$$+ K_x \Phi_{reg}(T_r L_y y_k + T_r L_u u(k-1)) + f_0 - s_d(k+1)\}, \qquad (4.15)$$

where Φ_{reg} and Γ_{reg} are the regular form of the matrices Φ_τ and Γ_τ, respectively.

4.3 Controller Design Procedure

The multirate output feedback-based frequency-shaped sliding mode control with nonswitching type of reaching law is designed to reduce the amplitude of vibration of the cantilever beam at resonance frequencies. The discrete-time system obtained by sampling the system in Eqs. (3.62) and (3.63) at a rate of $1/\tau$. The sampling interval is 0.01 s. The sensor output is sampled at 0.0025 s ($\Delta = \tau/N$) where $N = 4$. As shown in the frequency response of the smart structure model (Fig. 3.2), there are two modes of natural frequencies at 5.04 Hz (31.9 rad/s) and 32.84 Hz (204 rad/s), respectively. The frequency content of the optimal sliding mode is shaped using a frequency-dependent compensator presented in Eq. (3.34), such that a higher gain can be obtained at the natural frequencies. For nonswitching type of controller, the compensator dynamics are considered as

$$A_\omega = \begin{bmatrix} 0.9567 & -0.0861 \\ 0 & 0.4213 \end{bmatrix}; \qquad B_\omega = \begin{bmatrix} 0.6381 \\ 2.622 \end{bmatrix}$$

$$C_\omega = \begin{bmatrix} -0.6381 & -2.622 \end{bmatrix}; \qquad D_\omega = 15.66 \qquad (4.16)$$

With the above compensator, the augmented system (3.37) is constructed. And by solving the Riccati equation, the sliding gain K is obtained as

$$K = [-0.0039 \quad -0.1230 \ 0.1517 \ 0.1591 \ 0.0860].$$

The eigen values of the $(A_e - B_e K)$ are [0.9555   −0.2679   −0.0492 ± 0.8564i 0.0746]. Thus the augmented system in the sliding mode is stable. Using the above gain, the control law (4.15) is designed as

$$u(k) = 212.0168\{[-0.0038 \quad -0.0515]x_\omega(k)$$
$$-0.3249[0 \quad I_m](T_r L_y y_k + T_r L_u u(k-1))$$
$$+ [-0.8249 \ 0.3719 \ 0.5112 \ 0.3418](T_r L_y y_k + T_r L_u u(k-1))$$
$$+ d_0 - s_d(k+1)\}, \tag{4.17}$$

where,

$$L_y = \begin{bmatrix} -0.9976 & 3.7322 & -5.4729 & 3.7367 \\ -21.5126 & 65.5140 & -70.4864 & 26.4146 \\ -21.4627 & 53.9758 & -47.4234 & 14.9526 \\ -0.9904 & 0.0000 & -0.0000 & 0.0000 \end{bmatrix} ; L_u = \begin{bmatrix} -0.0000 \\ -0.0039 \\ -0.0024 \\ -0.0005 \end{bmatrix}$$

4.4 Controller Implementation

The experimental setup is shown in Fig. 3.3. Two piezoceramic patches are surface bonded at a distance of 5 mm from the fixed end of the beam. The patch bonded on the bottom surface acts as a sensor and the one on the top surface acts as an actuator. An excitation input is applied to the structure through another piezoceramic patch, which is bonded on the top surface at a distance of 370 mm from the fixed end. The dimensions and properties of the beam and piezoceramic patches are given in Tables 3.1 and 3.2. A piezoceramic of type SP-5H which is equivalent to NAVY TYPE VI, from Sparkler Ceramics Pvt. Ltd, India, is used in the experimental setup. The sensor output signal which is conditioned using a piezo-sensing system is given as analog input to the dSPACE1104 controller board. The control algorithm is developed using simulink software and implemented in real time on dSPACE1104 using RTW and dSPACE real-time interface tools. The simulink software is used to build control block diagrams and real time workshop is used to generate C code from the simulink model. The C code is then converted to target specific code by the real time interface and target language compiler supported by dSPACE1104. This code is then deployed on a rapid prototype hardware system to run the hardware in the loop simulation. The control signal generated from simulink is interfaced to a piezo-actuation system through the configurable analog input/output unit of dSPACE1104. The excitation signal is applied from the simulink environment through a DAC port of the dSPACE controller board.

4.5 Simulation and Experimental Results

To show the efficacy of the multirate output feedback-based frequency-shaped sliding mode control using nonswitching type of reaching law, the excitation frequency is changed from first mode to second mode after 10 s approximately. The simulation results and the experimental results with the nonswitching type FSSMC are shown in Figs. 4.1–4.3. In the case of nonswitching type of controller, the magnitude of vibration reduces to $\pm 5 \times 10^{-2}$ and $\pm 2.5 \times 10^{-2}$ V during the second vibratory mode excitement. Furthermore, to illustrate the effectiveness of the controller the frequency response of the system is studied when the beam is excited with a signal having first and second mode of frequencies (4.4). The response shows vibration reduction of approximately 8 dB at the first natural frequency and 18 dB at the second natural frequency.

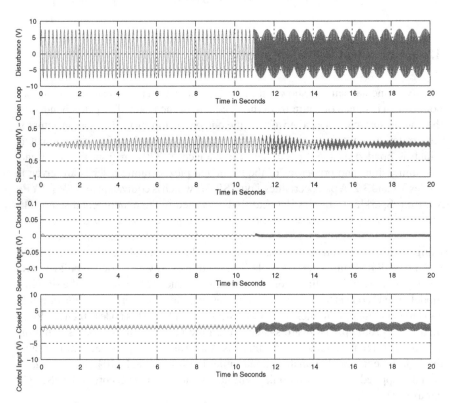

Fig. 4.1 Simulation results with nonswitching type multirate output feedback FSSMC when the excitation frequency is changed from first mode to second mode after 10 s approximately

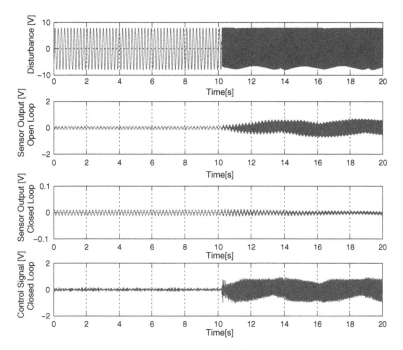

Fig. 4.2 Experimental results with nonswitching type multirate output FSSMC when the excitation frequency is changed from first mode to second mode after 10 s approximately

4.6 Comparison of Switching and Nonswitching Type Multirate Output Feedback FSSMC Performance

To study the effectiveness of switching type and nonswitching type controller, we compared the closed-loop responses. From the experimental results (Figs. 3.6 and 4.2), it is inferred that the vibration reduction during first mode excitation is ± 0.08 V in case of switching type and $\pm 5 \times 10^{-2}$ in nonswitching type control law, which shows that the vibration damping is more in switching type control law compared to nonswitching type control law. Moreover, it is also observed that the control efforts required for the nonswitching type controller is less and free of chattering compared to the case of switching type controller. Further, the frequency responses (Figs. 3.8 and 4.4) show the vibration reduction of 8 dB in switching type controller during first mode excitation which is more in comparison to 5 dB in case of non-switching type controller. Similar observations may be made for the second mode of excitation.

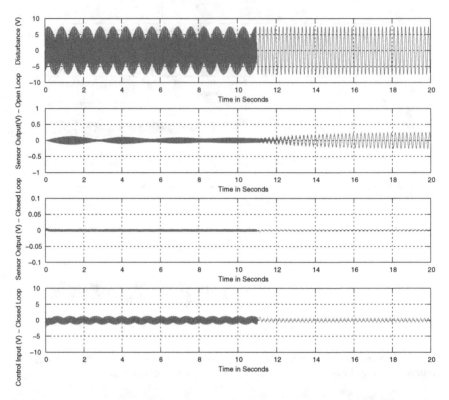

Fig. 4.3 Simulation results with nonswitching type multirate output feedback FSSMC when the excitation frequency is changed from second mode to first mode after 10 s approximately

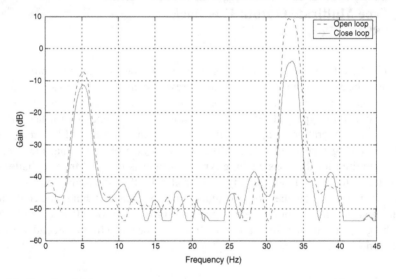

Fig. 4.4 Experimental frequency response

4.7 Conclusion

The chapter proposed multirate output feedback-based frequency-shaped sliding mode controller with nonswitching type reaching law. It can be inferred from the results that the control effort required in case of nonswitching type FSSMC is less and also free from the chattering compared to switching type of controller. Further, the results show that the controller continues to perform very well even after large changes in the excitation frequency.

References

1. Bartoszewicz A (1998) Discrete-time quasi sliding mode control strategies. IEEE Trans Ind Electron 45(04):633–637
2. Gao W, Wang Y, Homaifa A (1995) Discrete-time variable structure control system. IEEE Trans Ind Electron 42(02):117–122
3. Golo G, Milosavljevic C (2000) Robust discrete-time chattering free sliding mode control. Syst Control Lett 41:19–28
4. Mehta A, Bandyopadhyay B (2006) Multirate output feedback based frequency shaped sliding mode control. In: Proceedings of IEEE international conference on industrial technology (ICIT2006), Mumbai, India, pp 2658–2662
5. Mehta A, Bandyopadhyay B (2009) Frequency-shaped sliding mode control using output sampled measurements. IEEE Trans Ind Electron 56(01):28–35

Chapter 5
Reduced Order Observer Design Using Duality for Sliding Surface Design

Abstract This chapter gives the design of discrete-time reduced order observer design using duality between the discrete-time sliding surface design and discrete-time reduced order observer design. First, the duality between the discrete-time sliding surface design and discrete-time reduced or design of observer is explained. The advantage of the method is that the sliding surface and the observer are designed simultaneously. This method has been used to design Power System Stabilizer (PSS) of Single Machine Infinite Bus System (SMIB). Further, to show the efficacy of the method, a reduced order observer-based DSMC is designed for the laboratory experimental servo system and also implemented.

Keywords Discrete-time sliding mode control · Reduced order observer · Duality · Power system stabilizer · Industrial emulator servo system.

5.1 Introduction

In some situations, the multirate output feedback-based strategy may not perform well due to the presence of noise or sampling limitation. In these cases, the observer may be used for the output feedback-based controller design. A considerable amount of research has been carried out on observer and several authors have proposed several design methods for the observers. Observers use the plant input and output signals to generate an estimate of the plants state, which is then employed to close the control loop. The observer was first proposed and developed by Luenberger in the early 1960s of the last century [15–17]. Since then several authors have proposed the method for designing the observer [5, 22]. Among all the methods, the pole placement technique and the optimal observer design are widely used methods. Both the methods require the observer system to be transformed in to Gopinath's form [6]. But when the observer system is transformed to Gopinath's form, the original controllable canonical form is lost. Inoue et al. [7, 8] proposed a method for observer design in continuous-time domain using the duality concept in which the reduced order observer coefficients are directly obtained from the sliding surface design.

This chapter contains results from [19, 20] and additional results on the duality between the discrete-time sliding surface design and discrete-time reduced order observer design. This method has been used to design Power System Stabilizer (PSS)

© The Author(s) 2015   57
A. Mehta and B. Bandyopadhyay, *Frequency-Shaped and Observer-Based Discrete-time Sliding Mode Control,* SpringerBriefs in Applied Sciences and Technology, DOI 10.1007/978-81-322-2238-5_5

of Single Machine Infinite Bus System (SMIB) [18]. The Power System Stabilizers (PSS) has been used for long to enhance the stability and to improve the dynamic response of the power system [3]. The main concern of PSS is to extend the angular stability limits of the power system by providing supplemental damping to the oscillations of the synchronous machine rotors through the generator excitation [9]. This damping is provided by an electric torque applied to the rotor that is in phase with the speed variation. Once the oscillations are damped, the thermal limits of the tie lines in the system may be approached. This supplementary signal is useful during large power transfer and line outages [24]. Various control techniques for PSS design are proposed in the literature to improve the overall system performance. But the main problem with most of the controllers is that they need the entire state vector and thus the observer is necessary.

Further, the proposed reduced order observer-based DSMC is designed for the laboratory experimental servo system and also implemented to show the efficacy of the method. The servo system is an ideal for emulating control of modern industrial equipment such as spindle drives, turntables, conveyors, machine tools, and automated assembly machines [4].

5.2 Brief Review of Discrete-Time Reduced Order Observer

Consider the discrete-time linear time invariant system

$$x(k + 1) = \Phi_\tau x(k) + \Gamma_\tau u(k), \tag{5.1}$$
$$y(k) = Cx(k), \tag{5.2}$$

where $x \in \Re^n$ is the state variable, $\Phi_\tau \in \Re^{n \times n}$, $\Gamma_\tau \in \Re^{n \times m}$ is full rank, $u \in \Re^m$ is the control input, $C \in \Re^{p \times n}$ such that $C\Gamma_\tau$ is nonsingular, and $y \in \Re^p$ is the output. We assume that (Φ_τ, Γ_τ) is completely controllable and $m < n$.

Problem Statement: The problem is to design $(n-p)$th order observer to estimate the state variable $x(k)$ from the measurement $y(k)$ and input $u(k)$. The reduced order state observer is given by,

$$z(k + 1) = Dz(k) + Ey(k) + Fu(k), \tag{5.3}$$
$$\hat{x}(k) = Pz(k) + Vy(k), \tag{5.4}$$

where $\hat{x}(k)$ is an estimate of $x(k)$. The estimate $\hat{x}(k)$ converges to $x(k)$ if the coefficient matrices D, E, F, P, V satisfy the condition as given by

$$T\Phi_\tau - DT = EC, \quad F = T\Gamma_\tau, \tag{5.5}$$
$$PT + VC = I_n, \quad D \text{ is stable}, \tag{5.6}$$

where T is $(n - m) \times n$ matrix.

The observer design is to find the coefficient matrices satisfying the above conditions (5.5) *and* (5.6).

5.3 Duality to Discrete-Time Sliding Surface Design

Let the discrete-time sliding surface for the system in Eq. (5.1) is given by

$$s(k) = C_s x(k), \tag{5.7}$$

where s is of m dimensional vector and C_s is $\Re^{m \times n}$.

While selecting the matrix C_s of the siding surface for DSMC, the following conditions are to be satisfied:
(i)

$$\det(C_s \Gamma_\tau) \neq 0, \tag{5.8}$$

(ii) On the surface,

$$C_s x(k) = 0. \tag{5.9}$$

Without loss of generality, from condition (i), we may change the switching surface s of Eq. (5.7) to new switching surface

$$s' = (C_s \Gamma_\tau)^{-1} s, \tag{5.10}$$

then the sliding surface gain matrix becomes

$$C_s' = (C_s \Gamma_\tau)^{-1} C_s \tag{5.11}$$

and condition (i) becomes (i)$'$ as

$$C_s' \Gamma_\tau = I_m. \tag{5.12}$$

From this condition (i)$'$, we may say that there exist an $n \times (n - m)$ matrix W and $(n - m) \times n$ matrix U such that

$$\operatorname{rank} \begin{bmatrix} W & \Gamma_\tau \end{bmatrix} = n, \quad \text{and} \quad C_s' W = 0. \tag{5.13}$$

Hence, the condition (i)$'$ is rewritten as (i)$''$

$$\begin{bmatrix} U \\ C_s' \end{bmatrix} = \begin{bmatrix} W & \Gamma_\tau \end{bmatrix}^{-1} \tag{5.14}$$

or equivalently, from Eq. (5.14), we may write

$$WU + \Gamma_\tau C'_s = I_n. \tag{5.15}$$

This condition is equivalent to

$$\begin{bmatrix} U \\ C'_s \end{bmatrix} [W \ \Gamma_\tau] = \begin{bmatrix} UW & U\Gamma_\tau \\ C'_s W & C'_s \Gamma_\tau \end{bmatrix},$$

$$= \begin{bmatrix} I_{n-m} & 0 \\ 0 & I_m \end{bmatrix}. \tag{5.16}$$

Now, if we define state variables $x_1(k)$ and $x_2(k)$ as

$$\begin{bmatrix} x_1(k) \\ x_2(k) \end{bmatrix} = \begin{bmatrix} U \\ C'_s \end{bmatrix} x(k), \tag{5.17}$$

then the state on the sliding surface $C_s x(k) = 0$ also satisfies $C'_s x(k) = 0$. So from Eq. (5.17), we may write $x_2(k) = 0$. Multiplying both the sides of Eq. (5.17) with $[W \ \Gamma_\tau]$, we may write

$$[W \ \Gamma_\tau] \begin{bmatrix} x_1(k) \\ x_2(k) \end{bmatrix} = [W \ \Gamma_\tau] \begin{bmatrix} U \\ C'_s \end{bmatrix} x(k), \tag{5.18}$$

$$[W \ \Gamma_\tau] \begin{bmatrix} x_1(k) \\ x_2(k) \end{bmatrix} = x(k). \tag{5.19}$$

The system dynamics on the sliding surface is

$$x_1(k) = Ux(k)$$
$$x_1(k+1) = Ux(k+1)$$
$$= U(\Phi_\tau x(k) + \Gamma_\tau u(k))$$
$$= U\Phi_\tau (Wx_1(k) + \Gamma_\tau x_2(k)) + U\Gamma_\tau u(k)$$
$$= U\Phi Wx_1(k). \tag{5.20}$$

Hence, the condition (ii) is given by (ii)$'$

$$J = U\Phi_\tau W \quad \text{is stable.} \tag{5.21}$$

Lemma 5.1 *The condition* (ii)$'$ *is equivalent to the next condition* (ii)$''$

$$\Phi_\tau W - WJ = \Gamma_\tau L \quad \text{and} \quad J \text{ is stable.} \tag{5.22}$$

where $C'_s \Phi_\tau W = L$.

Table 5.1 Duality table

Observer coefficients	Sliding surface design coefficients
Φ_τ	Φ_τ^T
C	Γ^T
D	J^T
E	L^T
P	U^T
V	$C_s'^T$
T	W^T

Proof If condition (ii)$'$ holds, then substituting Eqs. (5.21) and (5.15) into the left-hand side of Eq. (5.22),

$$\Phi_\tau W - WJ = \Phi_\tau W - WU\Phi_\tau W$$
$$= \Phi_\tau W - (I_n - \Gamma_\tau C_s')\Phi_\tau W$$
$$= \Gamma_\tau C_s' \Phi_\tau W. \tag{5.23}$$

Substituting $C_s'\Phi_\tau W = L$, we get the right-hand side of (5.22). The designing prob-lem of sliding mode control law is to find the matrices W, U, C_s', J satisfying condi-tions (i)$''$ and (ii)$''$.

Transposing, (5.22) and (5.15),

$$W^T\Phi_\tau^T - J^T W^T = L^T\Gamma_\tau^T,$$
$$U^T W^T + s'^T\Gamma_\tau^T = I_n, \tag{5.24}$$
$$J^T \text{ is stable}$$

Comparing the condition (5.24) to (5.5), the duality of coefficients are obtained as shown in Duality Table 5.1. Using the duality, a design scheme for a discrete-time reduced order state observer is derived from the design of the discrete-time sliding surface and vice versa.

5.4 Design of Discrete-Time Reduced Order State Observer

It has been proved by Mita and Chen [21] for continuous-time system that the poles of the closed loop system on the sliding surface are equal to the zeros of the transfer function. Here, first we prove the same result for the discrete-time system by giving the Lemma 5.2 as given below.

Lemma 5.2 *The poles of the closed loop system on the switching surface, (the eigen values of J) are equal to the zeros of the pulse transfer function*

$$C_s(zI_n - \Phi_\tau)^{-1}\Gamma_\tau. \tag{5.25}$$

Proof The proof is given in the Appendix A.

The zeros of this pulse transfer function (5.25) are assigned to be stable by the matrix C_s. It can be obtained by solving the Riccati equation as given in the Theorem 1.

Theorem 1 *For a positive definite matrix $Q > 0$ and $\rho(\varepsilon) < 1$, if the matrix $\Phi_{\tau\varepsilon}$, solution of P_s from Riccati equation, and matrix C_s are defined as:*

$$\Phi_{\tau\varepsilon} = \Phi_\tau \varepsilon \tag{5.26}$$

$$P_s = Q + \Phi_{\tau\varepsilon}^T P_s \Phi_{\tau\varepsilon} + \Phi_{\tau\varepsilon}^T P_s \Gamma_\tau (R + \Gamma_\tau^T P_s \Gamma_\tau)^{-1} \Gamma_\tau^T P_s \Phi_{\tau\varepsilon} \tag{5.27}$$

$$C_s = \Gamma_\tau^T P_s \tag{5.28}$$

then all the zeros z_i $(i = 1, 2, \ldots, (n - m))$ of the pulse transfer function (5.25) are inside the $\vartheta(\varepsilon)$ in the z-plane where $\vartheta(\cdot)$ denotes the spectral radius and $\varepsilon < 1$ is a number.

To formulate the design method using duality, first we need to prove that the zeros of the pulse transfer function, $C_s(zI_n - \Phi_\tau)^{-1}\Gamma_\tau$ are equivalent to $C_s'(zI_n - \Phi_\tau)^{-1}\Gamma_\tau$.

Lemma 5.3 *The zeros of the pulse transfer function $C_s(zI_n - \Phi_\tau)^{-1}\Gamma_\tau$ are equivalent to $C_s'(zI_n - \Phi_\tau)^{-1}\Gamma_\tau$.*

Proof The zeros of the pulse transfer function $C_s(zI_n - \Phi_\tau)^{-1}\Gamma_\tau$ are equal to the roots of the equation, [10]

$$P(z) = \det \begin{bmatrix} zI_n - \Phi_\tau & \Gamma_\tau \\ C_s & 0 \end{bmatrix} = 0, \tag{5.29}$$

then we may write,

$$\begin{bmatrix} zI_n - \Phi_\tau & \Gamma_\tau \\ C_s & 0 \end{bmatrix} = \begin{bmatrix} zI_n - \Phi_\tau & 0 \\ 0 & I_r \end{bmatrix} \times \begin{bmatrix} I_n & 0 \\ C_s & I_r \end{bmatrix} \times \begin{bmatrix} I_n & (zI_n - \Phi_\tau)^{-1} \\ 0 & -C_s(zI_n - \Phi_\tau)^{-1}\Gamma_\tau \end{bmatrix}. \tag{5.30}$$

Then the determinant of Eq. (5.30) is

$$\det \begin{bmatrix} zI_n - \Phi_\tau & \Gamma_\tau \\ C_s & 0 \end{bmatrix} = \det(zI_n - \Phi_\tau) \det(-C_s(zI_n - \Phi_\tau)^{-1}\Gamma_\tau),$$

$$= \det(zI_n - \Phi_\tau) \det((C_s\Gamma_\tau)(-C_s'(zI_n - \Phi)^{-1}\Gamma_\tau)),$$

$$= \det(zI_n - \Phi_\tau) \det(C_s\Gamma_\tau) \det(-C_s'(zI_n - \Phi)^{-1}\Gamma_\tau),$$

$$= \det(zI_n - \Phi_\tau) \det \begin{bmatrix} zI_n - \Phi_\tau & \Gamma_\tau \\ C_s' & 0 \end{bmatrix}. \tag{5.31}$$

Hence, the roots of Eq. (5.29) are equal to the roots of

$$\det \begin{bmatrix} zI_n - \Phi_\tau & \Gamma_\tau \\ C_s' & 0 \end{bmatrix}. \tag{5.32}$$

Lemma 5.4 *The poles of the observer* (5.3) *(eigen values of D) are equal to zeros of the pulse transfer function*

$$C(zI_n - \Phi_\tau)^{-1}V. \tag{5.33}$$

Proof The proof is straightforward from Lemma 5.2 and the Duality Table 5.1.

From the Duality Table 5.1, we can obtain matrix V_R corresponding to s in sliding surface design and get the zeros of the transfer function $C(zI_n - \Phi_\tau)^{-1}V_R$ from the Theorem 1. From Eq. (5.28), we may write

$$C_s = \Gamma_\tau^T P_s. \tag{5.34}$$

Transposing,

$$C_s^T = P_s \Gamma_\tau \tag{5.35}$$

From Duality Table 5.1, the corresponding matrix is obtained as

$$V_R = P_s C^T \tag{5.36}$$

Similarly, from Eq. (5.11), we may write

$$C_s' = (C_s \Gamma_\tau)^{-1} C_s \tag{5.37}$$

Transposing,

$$C_s'^T = C_s^T (\Gamma_\tau^T C_s^T)^{-1} \tag{5.38}$$

From Duality Table 5.1, the corresponding matrix is obtained as

$$V = V_R (C V_R)^{-1} \tag{5.39}$$

Design procedure for discrete-time reduced order observer

1. Obtain the solution for P_s from (5.26) and (5.27) for any positive $\varepsilon < 1$ and obtain matrix V as

$$V_R = P_s C^T, \quad V = V_R (C V_R)^{-1}. \tag{5.40}$$

2. Find an $(n - m) \times n$ matrix T_0 satisfying

$$W_0 = \begin{bmatrix} T_0 \\ C \end{bmatrix}, \quad \text{rank}(W_0) = n, \tag{5.41}$$

and obtain the coefficients of observer T, D, E, F, P from the Duality Table 5.1 as

$$V_1 = T_0 V, \qquad T = T_0 - V_1 C, \tag{5.42}$$

$$D = T\Phi_\tau P, \qquad E = T\Phi_\tau V, \tag{5.43}$$

$$F = T\Gamma_\tau, \qquad P = W_0^{-1} \begin{bmatrix} I_{n-m} \\ 0 \end{bmatrix} \tag{5.44}$$

Theorem 2 *The coefficient matrices of reduced order observer obtained in step* (1) *and* (2) *above satisfy the observer conditions* (5.5).

Proof As the poles of D are in the unit circle then D is stable. Using Eq. (5.42),

$$W_0(PT + VC) = W_0 \left(W_0^{-1} \begin{bmatrix} I_{n-m} \\ 0 \end{bmatrix} [T_0 - T_0 VC] + VC \right)$$

$$= W_0 \tag{5.45}$$

Substituting, $D = T\Phi_\tau P$ and $PT + VC = I_n$ in to $T\Phi_\tau - DT$,

$$T\Phi_\tau - DT = T\Phi_\tau(I_n - PT) \tag{5.46}$$

$$= T\Phi_\tau VC \tag{5.47}$$

$$= EC \tag{5.48}$$

Remark 5.1 The matrix D depends on the selection of matrix T_0 which is not unique but by Theorem 2, poles of D are equal to the zeros of $C(zI_n - \Phi_\tau)^{-1}V$ and the corresponding pulse transfer function is independent of the selection of T_0 so are the poles of D.

5.5 Design of Power System Stabilizer of SMIB Using Proposed Duality

5.5.1 Power System Modeling

5.5.1.1 Small Signal Analysis of Single Machine Infinite Bus System

Consider a single machine infinite bus system shown in Fig. 5.1. For simplicity, it is assumed as synchronous machine represented by a model (neglecting damper windings both in d and q axes and also, the armature resistance). AVR and exciter are represented by a first-order transfer function as shown in Fig. 5.2 [23]

The algebraic equations of stator are

$$E_q' + x_d' i_d = v_q \tag{5.49}$$

$$-x_q i_q = v_d \tag{5.50}$$

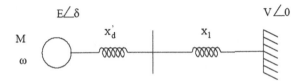

Fig. 5.1 Single line diagram of a single machine infinite bus system

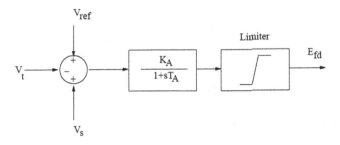

Fig. 5.2 Block diagram of excitation system

Linearizing Eqs. (5.49) and (5.50) as discussed in [23], the following equations are obtained

$$\Delta v_q = x'_d C_1 \Delta \delta + (1 + x'_d C_2) \Delta E'_q \tag{5.51}$$

$$\Delta v_q = -x_q C_3 \Delta \delta + x_q C_4 \Delta E'_q \tag{5.52}$$

5.5.1.2 Rotor Mechanical Equation and Torque–Angle Loop

The rotor mechanical equations are

$$\frac{d\delta}{dt} = \omega_B (S_m - S_{mo}) \tag{5.53}$$

$$2H \frac{dS_m}{dt} = -DS_m + T_m - T_e \tag{5.54}$$

$$T_e = [E'_q i_q - (x_q - x'_d) i_d i_q] \tag{5.55}$$

Linearizing Eq. (5.55), we get

$$\Delta T_e = [E'^{-(x_q - x'_d)}_{q0} \Delta i_q + i_{q0} \Delta E'_q - (x_q - x'_d) i_{q0} \Delta i_d] \tag{5.56}$$

We can express ΔT_e as,

$$\Delta T_e = K_1 \Delta \delta + K_2 \Delta E'_q \tag{5.57}$$

where

$$K_1 = E_{q0}C_3 - (x_q - x'_d)i_{q0}C_1 \tag{5.58}$$
$$K_2 = E_{q0}C_4 + i_{q0} - (x_q - x'_d)i_{q0}C_2 \tag{5.59}$$

Further linearizing Eqs. (5.53) and (5.54) and applying Laplace transform, we obtain [23]

$$\Delta\delta = \frac{\omega_B}{s}\Delta S_m = \frac{\omega_B}{s}\Delta\varpi \tag{5.60}$$

$$\Delta S_m = \frac{1}{2Hs}[\Delta T_m - \Delta T_e - D\Delta S_m] \tag{5.61}$$

5.5.1.3 Representation of Flux Decay

The equation for the field winding can be expressed as [23]

$$T'_{do}\frac{\mathrm{d}\Delta E'_q}{\mathrm{d}t} = E_{\mathrm{fd}} - E'_q + (x_d - x'_d)i_d \tag{5.62}$$

Linearizing Eq. (5.62), we have

$$T'_{do}\frac{\mathrm{d}\Delta E'_q}{\mathrm{d}t} = E_{\mathrm{fd}} - E'_q + (x_d - x'_d)(C_1\Delta\delta + C_2\Delta E'_q) \tag{5.63}$$

Taking Laplace transform of Eq. (5.63), we get

$$(1 + sT'_{do}K_3)\Delta E'_q = K_3\Delta E_{\mathrm{fd}} - K_3K_4\Delta\delta \tag{5.64}$$

where

$$K_3 = \frac{1}{[1 - (x_d - x'_d)C_2]} \tag{5.65}$$

$$K_4 = -(x_d - x'_d)C_1 \tag{5.66}$$

5.5.1.4 Representation of Excitation System

The block diagram of the excitation system is shown in Fig. 5.2. The linearized equations of the excitation system can be represented by omitting limiter. For the present analysis, we can ignore the auxiliary signal V_s. The perturbation in the terminal voltage V_t can be expressed as

$$\Delta V_t = \frac{v_{do}}{v_{to}}\Delta v_d + \frac{v_{qo}}{v_{to}}\Delta v_q \tag{5.67}$$

substituting for the values of Δv_d and Δv_q, we get

$$\Delta V_t = K_5 \Delta\delta + K_6 \Delta E'_q \tag{5.68}$$

where

$$K_5 = -\frac{v_{do}}{v_{to}} x_q C_3 + \frac{v_{qo}}{v_{to}} x'_d C_1 \tag{5.69}$$

$$K_6 = -\frac{v_{do}}{v_{to}} x_q C_4 + \frac{v_{qo}}{v_{to}} (1 + x'_d C_2) \tag{5.70}$$

The coefficients K_1 to K_6 are termed as Heffron-Phillips constants. They are dependent on machine operating conditions.

5.5.1.5 Computation of Heffron-Phillips Constants for Lossless Network

For $R_e = 0$, the expression for the constants K_1 to K_6 are simplified. As the armature resistance is already neglected, this refers to a lossless network on the stator side. The expressions of which are given below.

$$K_1 = \frac{E_b E_{qo} \cos\delta_0}{x_e + x_q} + \frac{(x_q - x'_d)}{x_e + x'_d} E_b i_{qo} \sin\delta_0 \tag{5.71}$$

$$K_2 = \frac{(x_e + x_q)}{x_e + x'_d} i_{q0} = \frac{E_b \sin\delta_0}{x_e + x'_d} \tag{5.72}$$

$$K_3 = \frac{(x_e + x'_d)}{x_d + x_e} \tag{5.73}$$

$$K_4 = \frac{(x_d - x'_d)}{x'_d + x_e} E_b \sin\delta_0 \tag{5.74}$$

$$K_5 = \frac{-x_q v_{do} E_b \cos\delta_0}{(x_e + x_q) v_{t0}} - \frac{-x_d v_{q0} E_b \sin\delta_0}{(x_e + x'_d) v_{t0}} \tag{5.75}$$

$$K_6 = \frac{x_e}{(x_e + x'_d)} \left(\frac{v_{q0}}{v_{t0}}\right) \tag{5.76}$$

In the above equations, the subscript '0' indicates the equilibrium value of the variable.

5.5.1.6 System Representation

The overall block diagram of the system, consisting of the representation of the rotor swing equations, flux decay, and excitation system is shown in Fig. 5.3. Here, the

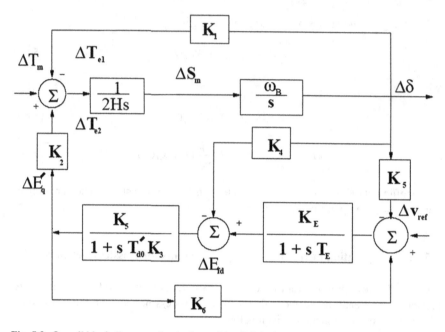

Fig. 5.3 Overall block diagram of a single machine infinite bus system

damping term (D) in the swing equations is neglected for convenience. From Fig. 5.3, the electrical torque compound ΔT_{e2} is related to $\Delta\delta$ by the following relation:

$$\Delta T_{e2}(s) = -\frac{K_2 K_4}{T'_{do}} F(s) \tag{5.77}$$

where

$$F(s) = \left[\frac{s + \frac{1}{T_e}(1 + \frac{K_5 K_E}{K_4})}{s^2 + s(\frac{1}{T_e} + \frac{1}{K_3 T'_{do}}) + (1 + K_3 K_6 K_E)/(K_3 T'_{do} T_E)} \right] \tag{5.78}$$

For static exciter, T_E is very small. If $T_E \approx 0$, $\frac{\Delta T_{e2}}{\Delta\delta}$ can be approximated as

$$\frac{\Delta T_{e2}}{\Delta\delta} \approx -\frac{K_2 K_4}{T'_{do}} \frac{(1 + \frac{K_5 K_E}{K_4})}{[s + (1 + K_3 K_6 K_E)/(K_3 T'_{do})]} \tag{5.79}$$

For large value of K_E, Eq. (5.79) can be further approximated as

$$\frac{\Delta T_{e2}}{\Delta\delta} \approx -\frac{K_2 K_4 K_E}{T'_{do} s + K_6 K_E} = -\frac{\frac{K_5 K_E}{K_4}}{T'_{do} s/(K_6 K_E) + 1} \tag{5.80}$$

The slip ΔS_m of machine is taken as output. The transfer function of the system is approximated as

$$\Delta S_m = \frac{1}{2Hs} \left[\Delta T_m - \left(\frac{\frac{K_5 K_E}{K_4}}{T'_{do} s / (K_6 K_E) + 1} + K_1 \right) \Delta \delta \right] \tag{5.81}$$

Taking ΔT_m small,

$$\Delta S_m = \frac{1}{2Hs} \left(\frac{\frac{K_5 K_E}{K_4}}{T'_{do} s / (K_6 K_E) + 1} + K_1 \right) \Delta \delta \tag{5.82}$$

5.5.1.7 State Space Model of Single Machine Infinite Bus System

The state space model of a SMIB power system, the block diagram for which is shown in Fig. 5.3 can be obtained using generator, transformer, network, and load flow data and is given below [3, 23]

$$\dot{x} = Ax + B(\Delta V_{\text{ref}} + \Delta V_s) \tag{5.83}$$

$$y = Cx \tag{5.84}$$

where x denotes the states of the machine and are given as $x = [S_m, \delta, E_{\text{fd}}, E'_q]$. Where S_m is machine slip, δ is machine shaft angular displacement in degrees, E_{fd} is generator field voltage in pu, and E'_q is voltage proportional to field flux linkages of machine in pu. The elements of the matrix A are dependent on the operating condition and is given as

$$A = \begin{bmatrix} 0 & \omega_B & 0 & 0 \\ -\frac{K_1}{2H} & \frac{-D}{2H} & \frac{-K_2}{2H} & 0 \\ \frac{-K_4}{T_{do}} & 0 & -\frac{1}{T'_{do}} & \frac{1}{T'_{do}} \\ \frac{-K_E K_5}{T_E} & 0 & -\frac{K_E}{K_6} & -\frac{1}{T_E} \end{bmatrix} \tag{5.85}$$

$$B = \begin{bmatrix} 0 & 0 & 0 & \frac{K_E}{T_E} \end{bmatrix} \tag{5.86}$$

$$C = \begin{bmatrix} 1 & 0 & 0 & 0 \end{bmatrix} \tag{5.87}$$

The mechanical damping term D, is included in the swing equation. The eigenvalues of the matrix should lie in LHP in the s plane for the system to be stable. The effect of various parameters (for example K_E and T_E) can be examined from the eigenvalue analysis.

5.5.2 Power System Stabilizer

It is well established that the fast acting exciters with high-gain AVR can contribute to oscillatory instability in power systems. This type of instability is characterized by low-frequency (0.2–3.0 Hz) oscillations which can persist (or even grow in magnitude) for no apparent reasons [23]. The major factors that contribute to the instability are

1. Loading of the generator or tie line
2. Power transfer capability of transmission lines
3. Power factor of the generator (leading power factor operation is more problematic than lagging power factor operation)
4. AVR gain

A cost efficient and satisfactory solution to the problem of oscillatory instability is to provide damping for generator rotor oscillations. This is conveniently done by providing Power System Stabilizers (PSS) which are supplementary controllers in the excitation systems. The signal V_s in Fig. 5.4 is the output from PSS which has input signal derived from the rotor speed, frequency, electrical power, or a combination of these variables. The objective of designing PSS is to provide additional damping torque without affecting the synchronizing torque at critical frequencies [3].

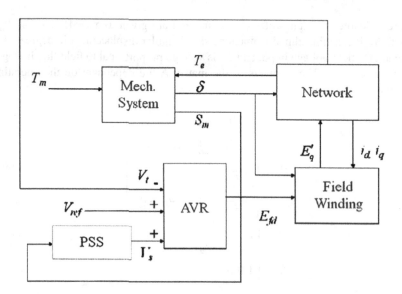

Fig. 5.4 Block diagram including excitation system and PSS

5.5.2.1 Basic Concept

The basic function of a power system stabilizer is to extend stability limits by modulating generator excitation to provide damping to the oscillation of synchronous
machine rotors relative to one another. The oscillations of concern typically occur
in the frequency range of approximately 0.2–3.0 Hz, and insufficient damping of
these oscillations may limit the ability to transmit power. To provide damping, the
stabilizer must produce a component of electrical torque, which is in phase with
the speed changes. The implementation details differ, depending upon the stabilizer
input signal employed.

Performance objectives
The main objective of providing PSS is to increase the power transfer in the network, which would otherwise be limited by the oscillatory instability. The PSS also
must function properly when system is subjected to large disturbances. PSS can
extend power transfer stability limits which are characterized by lightly damped
or spontaneously growing oscillations in the 0.2–3.0 Hz frequency range. This is
accomplished via excitation control to contribute damping to the system modes of
oscillations. Consequently, it is the stabilizers' ability to enhance damping under the
least stable conditions. Additional damping is primarily required under the conditions
of weak transmission and heavy load which may occur, while attempting to transmit
power over long transmission lines from the remote generating plants or relatively
weak tie between systems. Contingencies, such as line outage, often precipitate such
conditions. Hence, systems normally have adequate damping that can often benefit
from the stabilizers during such conditions.

5.5.2.2 Choice of Input Signal

The input signal for the PSSs in the system is a point of debate. The signals that have
been identified as valuable include deviations in the rotor speed ($\Delta\omega = \omega - \omega_0$),
the frequency (Δf), the electrical power (ΔP_e), and the accelerating power (ΔP_a).
Controllers based on speed deviation would ideally use a differential type of regulation and a high gain. Since this is rather impractical, the lead–lag PSS structure
is commonly used. However, one of the limitations of the speed-input PSS is that
it may excite torsional oscillatory modes [12, 23]. A power/speed ($\Delta P_e - \omega$, or
$\Delta P_a - \omega$) PSS design was proposed as a solution to the torsional interaction problem
suffered by the speed-input PSS [14]. The power signal used is the generator electrical power, which has high torsional attenuation. Due to this, the gain of the PSS may
be increased without the resultant loss of stability, which leads to greater oscillation
damping [23]. A frequency-input controller has been investigated and it has been
found that frequency is highly sensitive to the strength of the transmission system—
that is, more sensitive when the system is weaker—which may offset the controller
action on the electrical torque of the machine [12]. Other limitations include the

presence of sudden phase shifts following rapid transients and large signal noise induced by industrial loads [11]. On the other hand, the frequency signal is more sensitive to inter-area oscillations than the speed signal, and may contribute to better oscillation attenuation [12–14]. The use of a power signal as input, either the electrical power (ΔP_e) or the accelerating power ($\Delta P_a = P_{mech} - P_{elec}$), has also been considered due to its low level of torsional interaction. The ΔP_a signal is one of the two involved in the 4-loop AVR/PSS controller from [2], even though the tuning method related to this design approach is valid for other input signals. Although there are merits and demerits of each kind of choice of input used for PSS design, we choose rotor speed as input to illustrate PSS design using SMC technique. However, design can easily be adapted for other signals.

5.5.3 PSS Design for SMIB System

5.5.3.1 Linearization of Power System

The nonlinear differential equations governing the behavior power system can be linearized about a particular operating point to obtain a linear model which represents the small signal oscillatory response of a power system. The following parameters are used for simulation of the single machine infinite bus system model [23]: $H = 5\,\text{s}$, $T'_{do} = 6\,\text{s}$, $K_E = 100$, $T_E = 0.02\,\text{s}$ and $x_e = 0.2\,\text{pu}$. The slip of the machine is taken as output. A SIMULINK-based block diagram including all the nonlinear blocks has been used to generate the linear state space model of the system obtained as given below

$$A = \begin{bmatrix} 0 & 0.3 & 0 & -0.3 \\ 377 & 0 & 0 & 0 \\ 0 & -177.7 & -50 & -1731.4 \\ 0 & -9 & 2 & -1.2 \end{bmatrix}; B = \begin{bmatrix} 0 \\ 0 \\ 5000 \\ 0 \end{bmatrix} \tag{5.88}$$

$$C = \begin{bmatrix} 1 & 0 & 0 & 0 \end{bmatrix} \tag{5.89}$$

This linear continuous-time model is then discretized with the sampling time $\tau = 0.02\,\text{s}$ to obtain discrete-time state space model as

$$\Phi_\tau = \begin{bmatrix} 0.9812 & -0.0049 & -0.0000 & -0.0053 \\ 7.4925 & 0.9812 & -0.0000 & -0.0205 \\ -9.1184 & -1.9570 & 0.3257 & -21.0051 \\ -0.0795 & -0.0225 & 0.0029 & 0.9189 \end{bmatrix}; \Gamma_\tau = \begin{bmatrix} -0.0003 \\ -0.0007 \\ 61.5420 \\ 0.1701 \end{bmatrix} \tag{5.90}$$

5.5.3.2 PSS Design with Reduced Order Observer for Single Machine Connected To Infinite Bus

For the design of the reduced order observer-based PSS, first the desired locations of the poles are assigned for the $\varepsilon = 0.55$ and the matrix P_s is obtained from the Riccati equation. The matrix T_0 is taken as

$$T_0 = \begin{bmatrix} 0 & 1 & 0 & 0 \\ 1 & 0 & -1 & 1 \\ 0 & 0 & 0 & 1 \end{bmatrix} \tag{5.91}$$

and the reduced order observer is obtained as

$$z(k+1) = \begin{bmatrix} 0.9812 & 0.0000 & -0.0205 \\ 1.9345 & 0.3228 & 21.6012 \\ -0.0225 & -0.0029 & 0.9218 \end{bmatrix} z(k) + \begin{bmatrix} 7.4926 \\ 8.9934 \\ -0.0793 \end{bmatrix} y(k)$$

$$+ \begin{bmatrix} -0.0007 \\ -61.3719 \\ 0.1701 \end{bmatrix} u(k)$$

$$\hat{x}(k) = \begin{bmatrix} 0 & 0 & 0 \\ 1 & 0 & 0 \\ 0 & -1 & 1 \\ 0 & 0 & 1 \end{bmatrix} z(k) + \begin{bmatrix} 1.0000 \\ -0.0019 \\ -0.0000 \\ -0.0020 \end{bmatrix} y(k) \tag{5.92}$$

The control law is derived using the same sliding surface as

$$u(k) = -(C_s \Gamma_\tau)^{-1} [C_s \Phi_\tau x(k) - (1 - q\tau)s(k) + \rho\tau \operatorname{sgn}(s(k))] \tag{5.93}$$

5.5.3.3 Simulation with Nonlinear Model

The slip of the machine is taken as output to estimate the states by the designed reduced order observer. The control signal obtained by the state feedback control law and a limiter is added to V_{ref} signal. This signal is used to damp out the small signal disturbances via modulating the generator excitation. The disturbance considered here is a self-clearing fault which is cleared after 0.1 s. The limits of PSS output are taken as 0.1. Simulation results at different operating points (P_{g0}) are shown in Figs. 5.5, 5.6, 5.7, 5.8, 5.9, and 5.10. As shown in plots, the proposed PSS is able to damp out the oscillations in 3–4 s after clearing the fault for the active power in the range of, $P_{g0} = 0.5$ pu to $P_{g0} = 1.0$ pu with external line inductance of $x_e = 0.25$ pu to $x_e = 0.6$ pu.

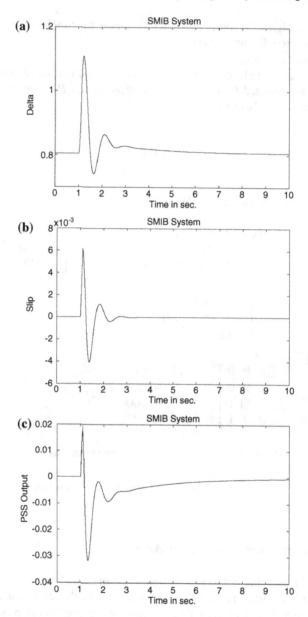

Fig. 5.5 **a** Delta. **b** Slip. **c** PSS Output response for $P_{gO} = 0.5\,\text{pu}$, $V_{\text{ref}} = 1.0\,\text{pu}$, $X_e = 0.25\,\text{pu}$

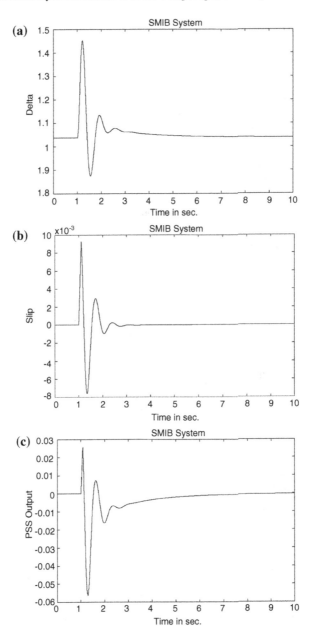

Fig. 5.6 **a** Delta. **b** Slip. **c** PSS Output response for $P_{gO} = 0.75$ pu, $V_{ref} = 1.0$ pu, $X_e = 0.25$ pu

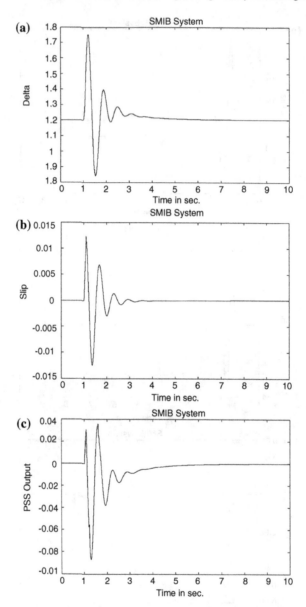

Fig. 5.7 a Delta. **b** Slip. **c** PSS Output response for $P_{gO} = 1.0$ pu, $V_{ref} = 1.0$ pu, $X_e = 0.25$ pu

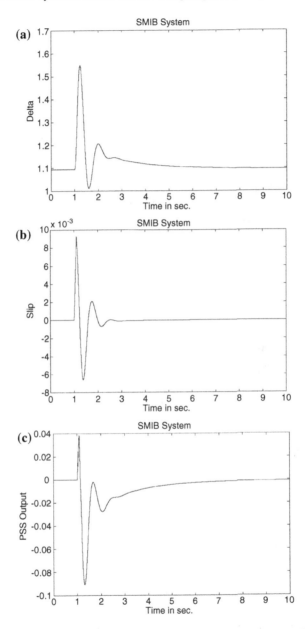

Fig. 5.8 **a** Delta. **b** Slip. **c** PSS Output response for $P_{gO} = 1.0$ pu, $V_{ref} = 1.0$ pu, $X_e = 0.35$ pu

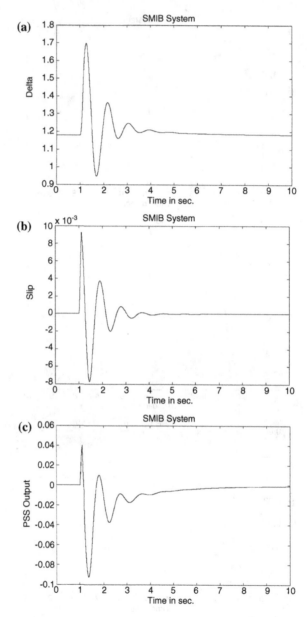

Fig. 5.9 **a** Delta. **b** Slip. **c** PSS Output response for $P_{gO} = 1.0\,\text{pu}$, $V_{\text{ref}} = 1.0\,\text{pu}$, $X_e = 0.45\,\text{pu}$

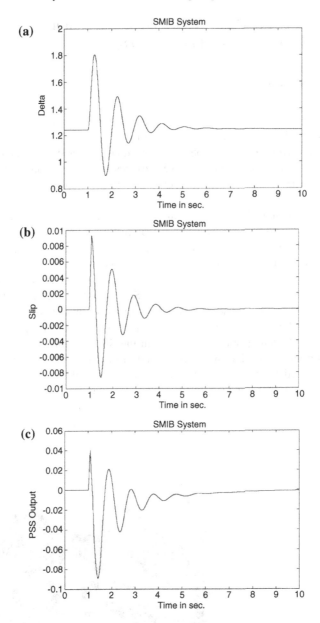

Fig. 5.10 **a** Delta. **b** Slip. **c** PSS Output response for $P_{gO} = 1.0\,\text{pu}$, $V_{\text{ref}} = 1.0\,\text{pu}$, $X_e = 0.6\,\text{pu}$

5.6 Reduced Order Observer Design Using Duality for the Industrial Emulator Servo System

5.6.1 System Description

The laboratory experimental servo system called Industrial Emulator Servo System shown in Fig. 5.11 is ideal for emulating control of modern industrial equipments such as spindle drives, turntables, conveyors, machine tools, and automated assembly machines [4]. A built-in secondary drive provides for programmable disturbance inputs and with the optional USER Executive software, the apparatus becomes a full MIMO test bed.

The electromechanical plant shown in Fig. 5.11 consists of the emulator mechanism, its actuator, and sensors. The design features brush less DC servo motors for both drive and disturbance generation, high resolution encoders, adjustable inertias, and changeable gear ratios. It consists of a drive motor (servo actuator) which is coupled via a timing belt to a drive disk with variable inertia. Another timing belt connects the drive disk to the speed reduction (SR) assembly while a third belt completes the drive train to the load disk. The load and drive disks have variable inertia which may be adjusted by moving (or removing) brass weights. Speed reduction is adjusted by interchangeable belt pulleys in the SR assembly. Backlash may be introduced through a mechanism incorporated in the SR assembly, and flexibility may be introduced by an elastic belt between the SR assembly and the drive disk.

A disturbance motor connects to the load disk via a 4:1 speed reduction and is used to emulate viscous friction and disturbances at the plant output. A brake below the load disk may be used to introduce coulomb friction.d The drive and disturbance

Fig. 5.11 The industrial emulator system

motors are electrically driven by servo amplifiers and power supplies in the controller box. Gear ratios can be changed via selection of the sizes of the upper and the lower pulleys in the SR assembly. The end-to-end gear ratio, i.e., gear ratio between load disk and drive disk is given by [4],

$$n_g = (N/n_{pl}) * (n_{pd}/n) \qquad (5.94)$$

where,
n = number of teeth on the drive disk pulley = 12
N = number of the teeth on the load disk pulley = 72
n_{pl} = number of teeth on the bottom pulley of the SR assembly
n_{pd} = number of teeth on the top pulley of the SR assembly

   High resolution incremental encoders are coupled directly to the drive and load disks to measure the incremental displacement of the motor and the load shaft. Each encoder has a resolution of 4,000 pulses per revolution. The encoders are routed through the controller box to interface directly with the DSP board via a gate array that converts their pulse signals to numerical values.
   The system option provides two analog output channels in the control box which are connected to two 16-bit DACs which physically reside on the real-time controller. Each analog output has the range of $+/-10\,\text{V}$ (-32768 to $+32767$ counts) with respect to the analog ground. The outputs on these DACs are updated by the real-time controller as a low-priority task.

5.6.2 Rigid Body Plant Model

From the Fig. 5.12, the overall drive train gear ratio, g_r, is such that $\theta_1 = g_r\theta_2$. From which, we may write

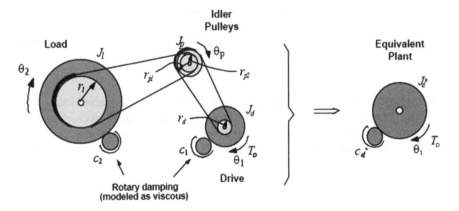

Fig. 5.12 Rigid body plant model of the industrial emulator system

$$g_r = \frac{r_l r_{p1}}{r_{p2} r_d} \qquad (5.95)$$

Referring the partial gear ratio between the idler pulley assembly and the drive disk g_r' as

$$g_r' = \frac{r_{p1}}{r_d}, \qquad (5.96)$$

we may write, $\theta_1 = g_r' \theta_p$.

Writing the torque balance equation as

$$T_l = g_r T_{dl}, \qquad (5.97)$$

where T_l is load torque acting on load and T_{dl} is drive torque acting on load.

Considering, $T_l = J_l \theta_1$, we have,

$$g_r T_{dl} = J_l \theta_1 = J_l g_r \dot{\theta}_d, \qquad (5.98)$$

implies,

$$T_{dl} = J_{lr} \dot{\theta}_d$$

where, $J_{lr} = g_r^{-2}$ is the load inertia reflected to the drive or input.

Same inverse squared scaling of inertia with speed ratio between elements holds. The drive inertia reflected to the load location, for example, that is $J_{dr} = J_d g_r^2$, then from Fig. 5.12, we may express

$$J_d^* = J_d + J_p g_r'^{-2} + J_l g_r'^{-2}, \qquad (5.99)$$

$$J_l^* = J_l + J_d g_r^2 + J_l \left(\frac{g_r}{g_r'} \right)^2, \qquad (5.100)$$

where J_d* and J_l* are the total inertias reflected to the drive and load, respectively. Similarly for the friction coefficients, it may be shown that

$$c_d^* = c_1 + c_2 g_r'^{-2}, \qquad (5.101)$$

$$c_l^* = c_2 + c_1 g_r^2 \qquad (5.102)$$

where c_d^* and c_l^* are the total reflected friction constants at the drive and load.

Note: For many applications involving servo drives, nonideal effects such as drive flexibility, backlash, static and kinetic friction, and other nonlinearities are sufficiently small that they may be neglected and the plant may be modeled as a simple rigid body.

Applying Newton's second law, we may express

$$J_d^* \ddot{\theta}_1 + C_d^* \dot{\theta}_1 = T_d, \tag{5.103}$$

$$J_l^* \ddot{\theta}_2 + C_l^* \dot{\theta}_2 = g_r T_d. \tag{5.104}$$

Using Laplace transform,

$$\frac{\theta_1}{T_d} = \frac{1}{s(J_d^* s + C_d^*)}, \tag{5.105}$$

$$\frac{\theta_2}{T_d} = \frac{g_r}{s(J_l^* s + C_l^*)} \tag{5.106}$$

When friction may be neglected, these reduce further to

$$J_d^* \ddot{\theta}_1 = T_d, \tag{5.107}$$

$$J_l^* \ddot{\theta}_2 = g_r T_d. \tag{5.108}$$

Using Laplace transform,

$$\frac{\theta_1}{T_d} = \frac{1}{J_d^* s^2}, \tag{5.109}$$

$$\frac{\theta_2}{T_d} = \frac{g_r}{J_l^* s^2)} \tag{5.110}$$

From Eq. (5.109), the continuous-time linearized model of the rigid body plant with the parameters given in [4] is obtained as

$$\begin{bmatrix} \dot{x}_1 \\ \dot{x}_2 \end{bmatrix} = \begin{bmatrix} 0 & 1 \\ 0 & -8.4344 \end{bmatrix} \begin{bmatrix} x_1 \\ x_2 \end{bmatrix} + \begin{bmatrix} 0 \\ 458.46 \end{bmatrix} u \tag{5.111}$$

$$y = \begin{bmatrix} 1 & 0 \end{bmatrix} \begin{bmatrix} x_1 \\ x_2 \end{bmatrix} \tag{5.112}$$

where, x_1 is the position of the drive motor and x_2 is the speed of the same.

5.6.3 Controller Data Board and Firmware

The real-time controller unit contains the digital signal processor (DSP)-based real-time controller, servo actuator interfaces, servo amplifiers, and auxiliary power supplies. The controller also interprets trajectory commands and supports such functions as data acquisition, trajectory generation, and system health and safety checks. Two optional auxiliary digital-to-analog converters (DAC's) provide for real-time analog signal measurement. The Executive program is the user's interface to the system and supports controller specification, trajectory definition, data acquisition, plotting, system execution commands, and more.

In addition to Executive program, Industrial Emulator Servo System also provides real-time Simulink environment. This allows to employ all the powerful design and analysis tools offered by Matlab and Simulink [4] while implementing real-time control and commanding the system through maneuvers using the block diagram environment of Simulink.

5.6.4 Simulation and Experimental Results with Switching Type Control Law

The discrete-time model of the system in Eq. (5.111) with sampling time $\tau = 0.01$ is obtained as

$$x(k + 1) = \Phi_\tau x(k) + \Gamma_\tau u(k) \tag{5.113}$$
$$y(k) = Cx(k) \tag{5.114}$$

where,

$$\Phi_\tau = \begin{bmatrix} 1 & 0.0096 \\ 0 & 0.9191 \end{bmatrix} \quad \Gamma_\tau = \begin{bmatrix} 0.0223 \\ 4.3966 \end{bmatrix} \quad C = \begin{bmatrix} 1 & 0 \end{bmatrix}$$

The state x_1 is available for measurement and x_2 need to be estimated by the proposed reduced order observer. Following the steps given in the design procedure, consider $\varepsilon = 0.7$ for solving the Riccati equation (5.27) and obtain sliding gain $C_s = [0.1153 \ 0.0021]$. Obtain the coefficients of observer as $D = 0.9190$, $E = -6.7958 \times 10^{-4}$, $F = 4.3964$, $P = [0 \ 1]^T$, and $V = [1 \ 0.0084]$ from step 2) of the design procedure. The switching type sliding mode control law given in Eq. (5.93) applied to the system is given as

$$u(k) = -(C_s\Gamma_\tau)^{-1}[C_s\Phi_\tau x(k) - (1 - q\tau)s(k) + \rho\tau\,\mathrm{sgn}(s(k))] \tag{5.115}$$

The control law (5.115) has two parameters ρ and q for tuning the response. From (2.38), ρ is directly proportional to the QSMB, and the system will overshoot when ρ is too large. On the other hand, large ρ could speed up transient response. From (5.115), $q\tau$ is required to be smaller than one, so q has to be smaller than $1/\tau$, but large q could speed up transient response. The parameters tuned for the control law are $q = 9$, $\rho = 0.15$.

The simulation and experimental results are shown in Figs. 5.13 and 5.14, respectively. It is observed from the results that the controller induces a chattering in the control input as well overshoot in the output response. So to improve the performance, we have designed the nonswitching type of sliding mode control law.

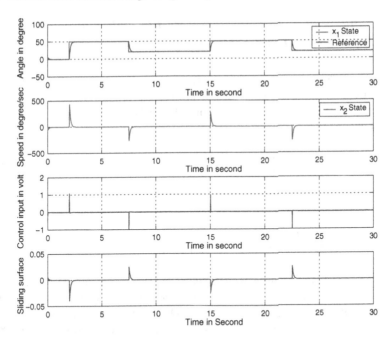

Fig. 5.13 Simulation results for the Emulator Servo System with the reduced order observer-based switching type of control law

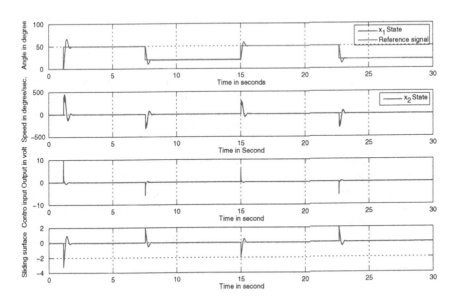

Fig. 5.14 Experimental results for the Emulator Servo System with the reduced order observer-based switching type of control law

5.6.5 Simulation and Experimental Results with Nonswitching Type Control Law

In order to design a Nonswitching type DSMC for the Industrial Emulator Servo System, we used Bartoszewicz's reaching law [1] as given in Sect. 2.4.1.

The discrete-time model of the system in Eq. (5.111) with sampling time $T_s = 0.006$ is obtained as

$$x(k + 1) = \Phi_\tau x(k) + \Gamma_\tau u(k),$$
$$y(k) = Cx(k), \tag{5.116}$$

where,

$$\Phi_\tau = \begin{bmatrix} 1 & 0.0059 \\ 0 & 0.9507 \end{bmatrix} \quad \Gamma_\tau = \begin{bmatrix} 0.0081 \\ 2.6823 \end{bmatrix} \quad C = \begin{bmatrix} 1 & 0 \end{bmatrix}.$$

The state x_1 is available for measurement and x_2 needs to be estimated by the proposed reduced order observer. Consider $\varepsilon = 0.7$ for solving the Riccati equation (5.27) and obtain sliding gain $C_s = [0.2845 \quad 0.0094]$. Obtain the coefficients of observer as $D = 0.9505, E = -9.1503 \times 10^{-4}, F = 2.6822, P = [0 \ 1]^T$, and $V = [1 \ 0.0185]$ from step (2) of the design procedure. The sliding mode control law applied to the system is given as

$$u(k) = -(36.2024)([0.2845 \quad 0.0106]x(k) + d_0 - s_d(k + 1)). \tag{5.117}$$

The simulation results and the experimental results are shown in Figs. 5.15 and 5.16, respectively, for a reference trajectory. It is observed that there is no overshoot in the position response and also no chattering. To verify the robustness property of the algorithm, a manual disturbance is applied at 23 s (Fig. 5.16). From the results, it is observed that the controller moves the states toward the surface when the disturbance is applied and the output once again tracks the reference trajectory.

Further, to verify the robustness and stability properties, a disturbance of $0.1\sin(0.7536k)$ is applied to the emulator system continuously through the input channel. Both the simulation and experimental results are observed and shown in Figs. 5.17 and 5.18, respectively. The results show that the position state is tracking the reference trajectory without any overshoot and steady state error. Moreover, the $\sigma(k)$ remains within a band of the sliding surface that is $|\sigma(k)| \leq \delta_d = 0.1$. This is also consistent with the theoretically calculated value from Eq. (2.45). This confirms the robustness and stability properties of the closed loop system under the bounded matched disturbance.

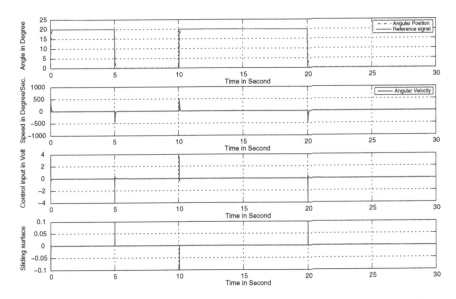

Fig. 5.15 Simulation results for the Emulator Servo System with the reduced order observer-based nonswitching type of control law

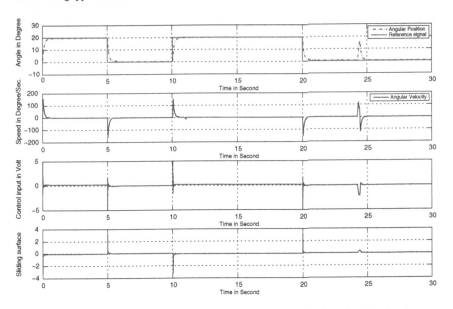

Fig. 5.16 Experimental results for the Emulator Servo System with the reduced order observer-based nonswitching type of control law

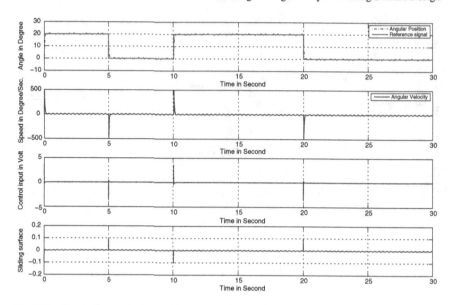

Fig. 5.17 Simulation results for the Emulator Servo System with the reduced order observer-based nonswitching type of control law for tracking a reference trajectory with matched disturbance

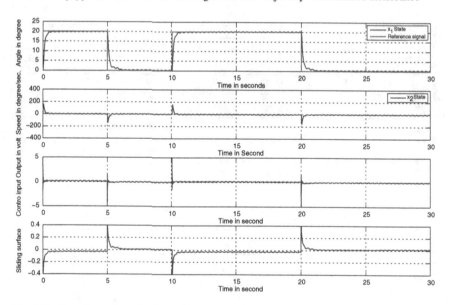

Fig. 5.18 Experimental results for the Emulator Servo System with the reduced order observer-based nonswitching type of control law for tracking a reference trajectory with matched disturbance

5.7 Conclusion

In this chapter, we proposed the duality between a discrete-time reduced order observer and the discrete-time sliding surface design. The discrete-time reduced order observer design method based on the solution of the Riccati equation is proposed. We designed the observer based on the proposed duality for PSS of SMIB system and it is found that the controller with the reduced order observer provides good damping enhancement for various operating points of the system. Simulation results from a nonlinear power system are given to demonstrate the applicability and effectiveness of the proposed approach. We also designed the sliding mode controller using the proposed reduced order observer design method for the Industrial servo system using switching type and nonswitching type control law to demonstrate the applicability and effectiveness of the proposed approach by simulation and experimental results.

References

1. Bartoszewicz A (1998) Discrete-time quasi sliding mode control strategies. IEEE Trans Ind Electron 45(04):633–637
2. Bourless H, Margotin T, Peres S (1997) Analysis and design of a robust coordinated AVR/PSS. In: IEEE-PES winter meeting, Anchorage
3. DeMello F, Concordia C (1969) Concepts of synchronous machine stability as affected by excitation control. IEEE Trans Power Appar Syst PAS 88:316–329
4. Educational Control Products USA (2004) Manual for model 220. Industrial Emulator/servo Trainer, ECP
5. Ellis G (2002) Observers for linear systems: a practical guide. Academic Press, San Diego
6. Gopinath G (1999) On the control of linear multiple input-output system. Bell Syst Technol J 50(01):50–55
7. Hamai Y, Inoue A, Deng M (2004) Duality between reduced-order observer and sliding mode controller and its application. In: Proceedings of SICE annual conference, Sapporo, vol 03
8. Inoue A, Hamai H, Deng M, Hirashima Y (2004) A design of an minimum-order observer by using the duality to sliding mode control law. Trans Inst Syst, Control Inf Eng 17(07):2145–2187
9. Jiang H, Cai H, Dorsey JF, Qu Z (1997) Towards a globally robust decentralized control for large scale power systems. IEEE Trans Control Syst Technol 05(03):309–319
10. Kailath T (1980) Linear systems. Prentice Hall, Englewood Cliffs
11. Kundur P (1993) Power system stability and control. McGraw-Hill Inc, New York
12. Larsen EV, Swann DA (1981) Applying power system stabilizers part-I: general concepts. IEEE Trans Power Appar Syst 100(06):3017–3024
13. Larsen EV, Swann DA (1981) Applying power system stabilizers part-III: practical considerations. IEEE Trans Power Appar Syst 100(06):3034–3048
14. Larsen EV, Swann DA (1981) Applying power system stabilizers part-II: performance objectives and tuning concepts. IEEE Trans Power Appar Syst 100(06):3025–3033
15. Luenberger DG (1966) Observers for multivariable systems. IEEE Trans Autom Control 11(02):190–197
16. Luenberger DG (1971) An introduction to observers. IEEE Trans Autom Control 16(06):596–602

17. Luenberger DG (1979) Introduction to dynamic systems: theory, models, and applications. Wiley, New York
18. Mehta A, Bandyopadhyay B (2007) Reduced order observer design for power system stabilizer using the duality to discrete time sliding surface design. In: Proceedings of 33rd annual conference of the IEEE industrial electronics society (IECON2007), Taipei, Taiwan, pp 908–914
19. Mehta AJ, Mehta HA (2012) Robust decentralized power system stabilizer for multi-machine power system using reduced order observer designed by the duality to sliding surface design. In: Proceedings of IEEE 12th international workshop on variable structure systems (VSS 2012), Mumbai, India, pp 142–148
20. Mehta A, Bandyopadhyay B, Inoue A (2010) Reduced-order observer design for servo system using duality to discrete-time sliding-surface design. IEEE Trans Ind Electron 57(11):3793–3800
21. Mita T, Chen YF (1990) Sliding mode control with application to the trajectory control of robot arm. Syst, Control Inf 34(01):50–55
22. O'Reilly J (1983) Observers for linear systems. Academic Press, New York
23. Padiyar KR (1996) Power system dynamics stability and control. Interline Publishing Private Ltd, Bangalore
24. Wang Y, Guo G, Hill DJ (1997) Robust decentralized nonlinear controller design for multi-machine power systems. Automatica 33(09):1725–1733

Appendix A

Consider the discrete-time system

$$x(k+1) = \Phi_\tau x(k) + \Gamma_\tau u(k) \tag{A.1}$$

In regular form [1], the system can be written as

$$x_1(k+1) = \Phi_{11} x_1(k) + \Phi_{12} x_2(k), \tag{A.2}$$

$$x_2(k+1) = \Phi_{21} x_1(k) + \Phi_{22} x_2(k) + \Gamma_{21} u(k), \tag{A.3}$$

where $x_1 \in \Re^{n-m}$ and $x_2 \in \Re^m$. B_{21} is nonsingular and $\Phi_{11}, \Phi_{12}, \Phi_{21}, \Phi_{22}$ are of appropriate dimensions. Define the sliding surface is

$$s(k) = C_s x(k), \tag{A.4}$$

where

$$C_s = \begin{bmatrix} C_{s1} & C_{s2} \end{bmatrix}, \tag{A.5}$$

where $C_{s2} \in \Re^{m \times m}$ is nonsingular. During ideal sliding

$$C_s x(k) = 0. \tag{A.6}$$

So

$$C_{s1} x_1 + C_{s2} x_2 = 0. \tag{A.7}$$

© The Author(s) 2015
A. Mehta and B. Bandyopadhyay, *Frequency-Shaped and Observer-Based Discrete-time Sliding Mode Control*, SpringerBriefs in Applied Sciences and Technology, DOI 10.1007/978-81-322-2238-5

Substituting for $x_2(k)$ in Eq. (A.2), then the ideal sliding motion is given by

$$x_1(k+1) = (\Phi_{11} - \Phi_{12}M)x_1(k), \tag{A.8}$$

where $M = C_{s2}^{-1}C_{s1}$. Comparing with Eq. (5.20), we may write

$$J = \Phi_{11} - \Phi_{12}M. \tag{A.9}$$

Now, the zeros of the pulse transfer function $C_s(zI - \Phi_\tau)^{-1}\Gamma_\tau$ are equal to the roots of the equation

$$P(z) = \det \begin{bmatrix} zI - \Phi_\tau & \Gamma_\tau \\ C_s & 0 \end{bmatrix} = 0, \tag{A.10}$$

called Rosenbrock's system matrix [2, 3]. Equivalently, it can be written as

$$\det(P(z)) = \det \begin{bmatrix} zI - \Phi_\tau & \Gamma_\tau \\ C_s & 0 \end{bmatrix} = 0 \tag{A.11}$$

$$= \det \begin{bmatrix} zI - \Phi_{11} & -\Phi_{12} & 0 \\ -\Phi_{21} & -\Phi_{22} & \Gamma_{21} \\ C_{s1} & C_{s2} & 0 \end{bmatrix} = 0.$$

As the Γ_{21} is assumed to be nonsingular

$$\det(P(z)) = 0 \Leftrightarrow \det \begin{bmatrix} zI - \Phi_{11} & -\Phi_{12} \\ C_{s1} & C_{s2} \end{bmatrix} = 0.$$

This is equivalent to

$$\det \left(\begin{bmatrix} I & \Phi_{12}C_{s2}^{-1} \\ 0 & I \end{bmatrix} \begin{bmatrix} zI - (\Phi_{11} - \Phi_{12}M) & 0 \\ 0 & C_{s2} \end{bmatrix} \begin{bmatrix} I & 0 \\ M & I \end{bmatrix} \right) = 0. \tag{A.12}$$

Equivalently,

$$\det \begin{bmatrix} zI - (\Phi_{11} - \Phi_{12}M) & 0 \\ 0 & C_{s2} \end{bmatrix} = 0. \tag{A.13}$$

As

$$\det \begin{bmatrix} I & \Phi_{12}C_{s2}^{-1} \\ 0 & I \end{bmatrix} = I \text{ and } \det \begin{bmatrix} I & 0 \\ M & I \end{bmatrix} = I, \tag{A.14}$$

the Eq. (A.12) is equivalent to

$$\det \begin{bmatrix} zI - (\Phi_{11} - \Phi_{12}M) & 0 \\ 0 & C_{s2} \end{bmatrix} = 0 \tag{A.15}$$

This means,

$$\det(P(z)) = 0 \Leftrightarrow \det(zI - (\Phi_{11} - \Phi_{12}M)) = 0 \qquad (A.16)$$

Therefore the invariant zeros of the $(\Phi_\tau, \Gamma_\tau, C_s)$ are the eigen values of the $(\Phi_{11} - \Phi_{12}M)$ which is equivalent to J in (5.21) i.e. the eigen values of the reduced order sliding motion.

References

1. Edwards C, Spurgeon SK (1998) Sliding mode control: theory and applications. Taylor and Francis, London
2. Kailath T (1980) Linear systems. Prentice Hall, Englewood Cliffs
3. Rosenbrock HH (1974) Computer-aided control system design. Academic Press, Orlando

Index

© The Author(s) 2015
A. Mehta and B. Bandyopadhyay, *Frequency-Shaped and Observer-Based Discrete-time Sliding Mode Control,* SpringerBriefs in Applied Sciences and Technology, DOI 10.1007/978-81-322-2238-5